THE SOCIAL LIFE OF BIOMETRICS

THE SOCIAL LIFE
OF BIOMETRICS

GEORGE C. GRINNELL

RUTGERS UNIVERSITY PRESS

New Brunswick, Camden, and Newark, New Jersey, and London

Library of Congress Cataloging-in-Publication Data

Names: Grinnell, George C., author.
Title: The social life of biometrics / George C. Grinnell.
Description: New Brunswick : Rutgers University Press, [2020] | Includes
    bibliographical references and index.
Identifiers: LCCN 2019030609 | ISBN 9781978809062 (paperback) |
    ISBN 9781978809079 (hardback) | ISBN 9781978809086 (epub) |
    ISBN 9781978809093 (mobi) | ISBN 9781978809109 (pdf)
Subjects: LCSH: Biometric identification—Social aspects.
Classification: LCC TK7882.B56 G75 2020 | DDC 303.48/34—dc23
LC record available at https://lccn.loc.gov/2019030609

A British Cataloging-in-Publication record for this book is available from the British
Library.

♾ The paper used in this publication meets the requirements of the American National
Standard for Information Sciences—Permanence of Paper for Printed Library
Materials, ANSI Z39.48-1992.

www.rutgersuniversitypress.org

Manufactured in the United States of America

For all those migrants who are dearly known, even if they are not identified

CONTENTS

CONTENTS

THE SOCIAL LIFE OF BIOMETRICS

INTRODUCTION

It is not easy to form an idea of the ground and what it knows. Consider the strip of land between Greece and Macedonia and the stories it has been told by the footfalls of migrants who fled war and insecurity in Syria following government repression of peaceful uprisings associated with the Arab Spring in 2011 and the subsequent conflict that displaced more than ten million people. What does this ground know of the aches, joy, and tears of those who walked upon it in search of a life that is livable? What does it know of the many reasons these individuals have for migrating to Europe and beyond, the pain and hope that propel them forward, or the desire to return home? How do the rumble of thousands upon thousands of feet and the accompanying echoes of laughter, fear, relief, unexpected friendships, and anguish forever change that earth, and how many times has it been changed in a similar manner in the past? What does the ground know of those who travel upon it in warmth and cold? What does it know of those who drowned in the Mediterranean or those who cannot undertake such a journey? What does it know of those who are stalled upon it, as so many were when several well-traveled tracts of land were transformed into refugee camps in 2016, no longer a path through Europe but now a site of detention and restriction?

We do not think much of the ground when considering moments of mass migration that began reaching crisis levels in 2015. Instead, we think of the individuals who move upon it and their stories and the attachments and heartaches that drive them. We think of the desires closely guarded and the dreams admitted not even to oneself. We think of what is left behind and what memories and troubles one carries along, not to mention what it takes to survive sometimes impossible situations. How strange that we conceive of individuals without the ground that transmits the weight of one individual to another. What does it mean to live without the ground that supports life, without the ground that makes it possible to move both together and singly?

This book is an attempt to chart what it means to think about individuals without ground, to dispossess, conceptually, individuals of the land that one walks upon and always shares with others. These are individuals who appear to walk upon no ground, naturally isolated from all others, and are known by an identity that estranges them from so much that is familiar and formative. This is a book about the emergence of a new type of ground, one that has intensified in the past two decades, but with a history of thought much older. My focus is on how biometrics produces a compelling idea of the body as a self-sufficient ground for identity. If biometrics was only a matter of verifying identity, matching a person to a record, there might be little to consider. But a central hypothesis of this book is that biometric practices transform how cultures understand the nature of identity and its parameters by fixing it so trenchantly to the physical body. Biometrics may authenticate identity, but its effects are not limited to such ends. This book considers what this technology does in the world, not only as a particular means of recording and verifying identity but as a mode of organizing reality and regulating human movement according to a set of priorities about how human beings are imagined to exist.

My operating premise is that biometrics refers to more than emergent technologies such as iris scanners and palm readers and to more than established methods such as fingerprinting and passport photography. Biometrics is not just a technology. It is a mode of thought. It is a mode of thought that conditions how we encounter identity, regulate access, and understand human mobility. Biometrics is a way of recognizing the individual and a way of establishing the ground for existence in a particular concept of the body and its limits. If it is an imaging technology that can identify individuals in new ways, such modes of seeing also constitute reality by noticing only some of what makes a life possible.

The Social Life of Biometrics is an attempt to document and theorize how some of the most powerful effects of biometrics exceed rituals of identity verification and amount, instead, to a mode of thinking about how we exist in the world individually and with others. The world is shaped by how we see it, as well as what we can and cannot imagine about it, and biometric thought refers to a social practice of framing concerns about identity, security, mobility, the physical body, and borders, among other related matters, that establish and condition how we live with and apart from one another. Biometrics has a social life, and this book is an attempt to understand what this involves, what it has done and continues to do, as well as how it functions.

BIOMETRIC FANTASIES

For a long time, practices of identification were informal and largely governed by face-to-face knowledge. The arrival of strangers was relatively uncommon, and individuals could learn essential information about another in the course of a simple conversation. This was a world in which identification revolved around the direct recognition of one another's shared humanity and a willingness to trust that someone was who they said they were. Such a time seems impossibly idealistic to our present, in which people routinely traverse the globe and social life has been transformed by a global economy as well as by legacies of colonial expansion, warfare, telecommunications, and ongoing migrations. Could such an uncomplicated time ever exist again? Like so many fantasies, the fantasy of face-to-face relations captures a sense of what once was, even and especially if it never actually existed. It is an alluring simplicity that biometrics takes advantage of when it reduces the complexity of identification to a scene of punctual face-to-face identification using advanced technology. To see the truth of another by looking into one's face is to embark on a path of evading and avoiding so much that explains who one is, how one exists, and what one needs to exist in the world.

Biometric thought asserts that identification is necessary in a dangerous world: one can no longer afford to live in the past. What does such an opposition prepare one to do, I wonder, positioning as it does our modern age as necessarily disappointing when set against a perhaps ideal world that is too old-fashioned to be practical? Weaponizing disappointment and disenchantment as necessary in the name of sober maturity and security, biometric thought prepares one to consent to a vast array of measures almost because the measures are undesirable. Perhaps there are good reasons for petulance rather than sensible maturity during times when biometric thought too often confirms that threats are almost always foreign and alien to a population, unrelated to the workings of politics and economics, and that safety is reliably ensured via discrete forms of identification.

No community was ever quite so immediate and intimate as the old-fashioned fantasy suggests, and perhaps no mode of recognizing others has ever worked without the intercession of social norms that make some strangers more strange than others. One need only think of the forms of division and exclusion that operate so powerfully in workplaces and schools to know how effectively intimate environments can be structured by judgments about what is proper or desirable. One can tend to forget that even small-scale communities are encumbered by mediating social forces that define how individuals recognize one another. Yet the lure of such a world is powerful, perhaps because it offers an antidote to the impersonality of modern modes of identification that increasingly rely on documents and social measures that remain foreign to our sense of everyday life. Curiously, biometrics is both a symptom of that official culture and an effort to breathe life into fantasies of human immediacy by insisting that one's physical presence is the truest expression of identity.

In this sense, *The Social Life of Biometrics* seeks to understand the social and cultural impact of a fundamental confusion between a technical measure of identification and the desires that both structure it and are emboldened by it. One might imagine that a discrete act of biometric confirmation of one's legal or national identity should not, perhaps, bear upon what one sees or looks for in the face of another. It should not impact how one understands human migration at the beginning of the twenty-first century or what it meant two hundred years ago. It should not help to normalize inhospitable measures or prompt discussions of ethics. It should not transform how we think of the body as a container of one's existence or make easy the disposal of social relations and a natural environment, as if those had little bearing on existence. It should not initiate reflections on democracy and sovereignty in the wake of global capitalism. Biometric identification should not make possible these and so many other considerations that mediate how we think and live in a world with others. But it does—often by appearing to be scientifically neutral and indifferent to the complex social matters it is asked to address. Biometric techniques promise only to confirm presented biological credentials with those already recorded in a database. Yet biometric applications are often made to do much more than this amid promises, for example, to secure the nation from dangerous outsiders, to presume that some individuals require scrutiny and not others, or to insist that some populations are especially suspect. The cultural life of biometrics can indeed make plain the social prejudice of those who grab hold of otherwise neutral techniques, but I have in mind a different critique than this. The social life of biometrics entails what I call biometric thought, a knowledge that draws upon and collects together a number of ways of framing social matters and ways of thinking and acting on the basis of biometric assumptions. Biometric identification depends upon reducing background static so as to only "see" that which is relevant for making a one-to-one identity match. Biometric thought generalizes this act of not seeing anything more than the individual and elevates it to a widespread form of social recognition that organizes how cultures and individuals come to perceive identity and one's relation to others as well as to oneself.

This book does not wax nostalgic about the time before biometric measures were used to assess human identity, about the simple modes of assessing individuals to discern who each one is. Nor does it lament that advanced biometrics marks the advent of a new era of widespread surveillance that has "become much more flexible and mobile, seeping and spreading into many life areas where once it had only marginal sway" (Bauman and Lyon 3). Indeed, a key premise of this book is that the ideas and concepts collected under the term *biometrics* are much older than is often assumed, even if many of the technical procedures associated with it seem unprecedented. *The Social Life of Biometrics* considers the emergence of our biometric present as well as its history and precedents and what kinds of effects it has in the world when it seeks to produce and insist upon an understanding of identification that can only see the biological person.

This approach contends that our present understanding of biometrics fails to grasp the ways in which practices of verifying identity mediate numerous pressing social concerns. *Biometrics* refers to a range of procedures that collect, convert, and verify unique biological data acquired by examining elements such as fingerprints, an iris, the structure of a face, the sound of a voice, or an identifiable gait. Biometric data can be included in passports and police files as well as in personal devices such as cell phones and computers; the scope of access control provided by biometrics ranges from the official to the ordinary. These procedures of identity verification, I argue, are premised on a number of social assumptions regarding identity and the nature of the world and come to condition how one exists and understands oneself and one's identity, profoundly affecting what one can imagine and know about the world and how one exists with others in it. To the extent that biometric thought is not reducible to biometric technology, it issues from a number of social and historical sources. Biometric thought is the product of government, commercial, and industry procedures as well as the cultural ideas and assumptions that inform the technology and its existence in the world.

What ideas and concepts are structured by this view of biometrics and its social existence? How does biometrics condition and rely upon particular modes of understanding, such as the assumption that a body is legible and composed of a series of discrete markers of identity? What does the social life of biometrics tell us about how identity and existence are presently understood? How does biometric thought perceive the world, and what is made more or less conceivable? What are the effects of biometric thought, and who is affected by it? Can it make some futures more or less possible on the basis of the recognitions it produces? Biometrics promises a great deal, including securing national borders and catching terrorists as well as protecting individuals from identity theft and unveiling strangers. Its greatest promises may speak to the presumed physical essence of how we exist in the world: each of us a master of the geography of the flesh, or perhaps its prisoner. Always on the threshold of becoming ubiquitous but perhaps never quite able to deliver on those lofty promises, dreams of a biometric future abound, and as with any dream, we can learn a great deal about our world by examining how its fantasies operate.

I study the social life of biometrics in a number of ways, starting with a series of case studies of discrete human interactions with biometric practices and assumptions. Subsequent analysis and elaboration are illustrated by fiction, art, politics, histories of technology, philosophy, poetry, journalism, and academic scholarship. I turn to such varied archives in order to reflect biometric thought's presence in so many locations of social and cultural life and to document the ways in which it has come to be seen as the answer to so many more social and political challenges than it can possibly address.

BIOMETRIC OPERATIONS AND EFFECTS

The first recorded use of the term *biometrics*, in the sense we presently expect, may have been as recent as 1982. The Oxford English Dictionary (OED) offers the following example from *American Banker* (Mar. 15, 1982): "Scientists who are developing ways to measure these unique biological traits have created a new technology called biometrics that promises some useful tools for banking operations."

The National Research Council in the United States defines *biometrics* as "the automated recognition of individuals based on their behavioral and biological characteristics" (15). Recognition involves different technical modes of matching. For example, whether at the border or when using one's face to unlock a phone, biometric technology matches the individual to previously recorded electronic representations by reading the unique contours of a fingerprint or the pattern of blood vessels visible in an iris scan or the overall composition of a face. This is known as one-to-one matching. Alternatively, identity might be determined by comparing one image to many others stored in a database to determine whether a match can be made based on statistically significant similarities. This is known as matching one to many. Automated recognition is a science of probabilities, not absolutes, so errors are possible, as are challenges with individuals whose various features or physical characteristics may not be as legible as the technology expects. It is a science not yet as perfect in execution as many cultural representations would have us believe. The association between security, identity, and the fixed biology of the individual is a departure from an older sense of biometrics, which understood the term to refer to a statistical analysis of living organisms, especially concerned with the duration of life and hereditary matters. Key to this older sense of biometrics is the unique application of a stable science of statistics to the unstable and unpredictable world of living creatures; OED offers this example from 1923 (W. M. Feldman, *Biomathematics*): "Biometrics is the application of modern statistical methods to the measurements of biological (variable) objects."

What is it about the sheer physicality of the body, about its tendency to be more real than anything else in the world, that gives biometrics its present credibility? Modern biometrics discovers the body to be something sure, stable, and foolproof, and it is hard not to suspect that the body has become the foil we need it to be in a world of uncertain threats. I highlight this because it begins to capture what we might think of as the creativity of the social life of biometrics. The idea that the body is durable and reliable has not always been held by this science. At the beginning of the twentieth century, biometry found that the vicissitudes of life and the degradations of the flesh could best be addressed with the broad precision of actuarial analysis concerned with life expectancy and mortality rates. Moreover, that the history of biometrics has developed two very different views of the body shows the extent to which the body itself has a history and a social and cultural existence. It is not a stable ground but instead something that comes to be identified in particular ways and according to given assumptions. Biometric thought is part of that history, producing understandings of the body that affect how people live.

This book offers a social and cultural perspective on biometrics in order to consider how it functions and what sorts of effects it has produced and what it might yet do. Are there modes of living with and against biometric impulses that have ignored or neglected expressions of biometric thought that might be worth hearing in the present? I am not convinced that biometric efforts to record and track individuals are particularly new, despite the shine that comes with the most inno-vative and newest technology. Fingerprinting is more than a century old. Facial recognition was a staple of European culture two hundred years ago, albeit in a decidedly low-tech manifestation that involved consulting pages and pages of por-trait profiles and assessing individual facial features for clues to reveal the inner life of a person. Have the assumptions that structured particular means of assess-ing the identity of strangers changed as much as the technologies seem to have changed? What one sees in the face of another and how a given culture chooses to understand the stakes of that query reveal a great deal about the social ambi-tions of biometric thought and its capacity to condition how one thinks.

Biometrics reinforces a particular idea of the natural self and its markings (fin-gerprints, facial features, gait, voice) even as it transforms that self into its oppo-site in the form of binary code. But even this transformation of the individual may not be especially new. Schools, universities, workplaces, and banks have long transformed individuals into data by tracking and scoring them on the basis of particular indicators. Why might we feel that more is at stake when biometric tech-nology lays hold of the body and seeks to render a representation of it in the form of data? Is it because one's physical existence is so closely associated with an idealized understanding of oneself? Perhaps. But if that is the case, I am struck that a concern with having the body transformed into data does more than just cause worry about biometric privacy and data security. I suspect that this con-cern also affirms the idea that I am my body. It may sound strange to consider the possibility that I am more than my body, but given a social world of attachments to others as well as all sorts of physical, emotional, and psychological supports that are beyond my physical existence that I nonetheless depend upon to be who I am, perhaps it is not too odd. Biometrics paradoxically reassures us of our indi-viduality precisely because it posits a physical essence that it can then transform into data. Perhaps fears of a dystopian biometric future reflect a familiar fantasy that one presently has an identity that one fully possesses. It is just that biomet-ric thought encourages one to perceive that identity as something both attached to my body but also something that can be stolen! Rather than fearing a future in which individuality is replaced by binary code—an anxiety that takes a great many logical leaps and seems unwilling to confront the generalized conditions of medi-ation and representation that have always defined human social existence—what I propose here is an attempt to understand biometrics as an indicator of the extent to which individual identity always involves social as well as personal realities.

One need only think about people who do not feel at home in their bodies to begin to recognize that a natural connection between oneself and one's body is a social construction. The examples are many: those who feel their gender does not match their body; those who do not wish to be defined by so-called physical dis-abilities; those whose capacity to empathize with others is particularly strong and whose emotional walls may be especially porous such that their experience of self-hood does not assume the presence of bodily borders between individuals; those whose daily life is shaped by racism and the experience of varying degrees of dispossession because social norms force them to see themselves, at least in part, through the prejudice of others. So, the idea that the body is a marker of the self may be a particular way of confirming individuality, but it may not be real, natural, or experienced in the same way by everybody. How one lives within his or her or their body may be an alienating experience rather than the most natural thing in the world.

This consideration of how biometrics is attached to an idea of the body exposes several assumptions that I will examine in this book. Does biometric thought distort a naturally existing body, or does it participate in a social process that produces an idea of the body that guides ways of acting and living in the world? Such a query gets to the heart of how biometrics may lead to entire modes of thought, not just punctual measures of identification. Biometrics has changed how we think about migration, individual identity, and one's attachments to a life-sustaining world. Even this brief consideration of the history of the word *biometrics* shows that the term itself is far from settled and contains a number of possible trajectories of influence. A founding assumption of this book is that biometric thought is at once consequential in its operations and effects and also not absolute or unchanging.

Biometric thought is attached to a range of concepts that it supports and which make its operations possible. It may not ever be solely responsible for the work it does in the world. The ideas about identity and human movement that biometrics sustains exist outside of its operations and ambitions. They are general concepts and ideals with rich histories and long-standing assumptions, and I am not trying to suggest that biometrics alone invents the self as a physical entity, for example. But certain concepts are especially concentrated and redirected under the influence of biometric thought. As a social practice that involves a set of ritualized technical procedures as well as aspirations, ideals, and assumptions, biometric knowing knits together disparate concerns with human migration, hospitality, territory, strangers, social perception and affect, security, aesthetics, law, physical existence, anonymity, knowledge, self-possession, papered and paperless identity—to name but a few of the domains enlivened by the social dimensions of biometrics.

Biometric thought involves more than biometric procedures. By the same token, biometric procedures are not limited to its application at the border. Biometrics is deployed to authenticate identity and police access across a range of real and virtual territories, including online banking, airports, personal computers, and cell phones. Yet it may be at the border where the stakes of authentication are most consequential, most visible, and where biometrics mediates the greatest number of social and cultural concepts and most profoundly serves to condition how one thinks about identity. This does not mean that what I have to say here applies only to biometrics at the border; however, a primary force of biometric thought needs to be understood as a means of regulating access via identification, and the normalizing force of such operations is creating ways of thinking about life on this planet that are undeniably quickened by the desire to regulate human movement. I wonder if there could ever be an account of measures of identification and access control that is not also an account of regulating human movement across borders, given how tightly these different matters have been braided together.

In his history of European measures of identification, Valentin Groebner maps several distinct ideas of identity, noting that it can "refer to membership in a collective" as readily as it can refer to an "internalized physical experience as a core of the self" (26). It makes little sense to speak of identity without also speaking about its temporality and the shifting understandings of the term as well as the different historical contexts in which one might speak of identity. For example, to speak of identity in the context of decolonization is to address a feature of individual and collective experience that has immediate social and political implications. In the context of digital activism by Anonymous, identity becomes an object of principled opposition and a legal liability. In opposition to the conceptual flexibility associated with the term *identity*, Groebner prefers to speak about identification, not identities, "because identification is always a process that involves more than one person. It seems to me that the various agents and authorities involved in naming and keeping individuals distinct cannot be brought into view otherwise, for they are always present" (27). I share his sense that it is necessary to chart the ways in which identity is conditioned by forces outside the individual, whether these issue from social institutions that produce identity documents encoded with biometric information, from political leaders reassuring their population that refugees will be screened biometrically, or from popular culture and its assumptions about those who live without identity papers. When it comes to thinking of identity and identification, there is always more to consider than the presence of just one person.

One of the primary effects of the social life of biometrics is to define the terms by which individuals can be recognized. In order to overcome the evasiveness of identity, biometrics produces a notion of the physical body as the ground for formal measures of identification. In place of the complexity of a social experience of identity, biometrics asserts that one's body, not a set of concepts or social norms, is the basis for who one is. It offers a mode of social recognition that begins as a particular way of seeing the body. If verifying one's identity becomes the basis for normal social existence, then that directly impacts how we conceptualize shared human existence and the differential right to have one's life respected. For example, biometrics may render invisible individuals who work without legitimate documentation so that they may not access social services or avail themselves of protections against exploitation. In another sense, biometric thought makes such individuals all too visible by identifying entire populations—races sometimes, but also industries, such as hotel workers or kitchen staff—as "illegals." Such designations forget the humanity of those they claim to see by reducing disparate individuals to a single attribute that says nothing about who they are but a great deal about the society that wishes to separate them from their human dignity.

If biometrics was a radically new way of seeing the body or signaled the arrival of a new age of absolute surveillance, there would be little to say in response to such a certain future. However, to speak of the effects and the consequences of biometric thought is not to designate a fixed field where only specific outcomes are possible, even if biometrics may aspire to constrain risk and difference. Sometimes the effects of biometric thought can be surprising and even contradictory.

Consider, for example, the questions provoked by the court case brought by the FBI against Apple in the wake of the mass shooting perpetrated by Syed Farook and Tashfeen Malik in 2015. The FBI wanted to access Farook's cell phone, and the court ordered Apple to unlock it. Had Farook used biometric encryption, it could have been unlocked readily, if somewhat morbidly, by using his fingerprint postmortem. Biometric security often cannot distinguish a living from a deceased fingerprint, and thus this case arose only because Farook did not use what many would consider the most technologically sophisticated means of controlling access to his phone. When a suspect is in custody in many places in the United States, moreover, police can compel an individual to activate a phone's biometric pass code, such as Face ID, but cannot similarly demand a numerical pass code (Waddell). The body speaks, but its speech is not protected under the right to avoid self-incrimination. The instrumentalization of the body as if it were a key runs counter to a biometric logic that suggests biometric identification guarantees a one-to-one relationship in which I am my body and vice versa. So, which is it? Do I coincide with my body, or is my body an instrument of sorts that can be used as if it were a tool, part of me but also not me? If my biometric profile is an indicator of my identity, does this make the flesh and blood that sustain my life more or less a part of my identity? Is this mode of perceiving the body and the person sufficient, or does it conceal other important factors that ought to be considered as well? This book tries to pry open these sorts of fissures in order to ask how biometric thought conceives of the individual and how its conceptions and assumptions may produce, rather than reflect, the very self they encounter.

These matters are all the more complex when we note that legal rights to privacy and personal information may often not obtain in instances most likely to be policed by biometrics, such as crossing a border. Entry into a country may depend on complying with a border officer's demands to access personal information, particularly when a noncitizen will simply be refused entry for failing to comply. What the law says and how it interprets biometric selfhood is, in short, far from the last word on biometrics, especially if one has no legal standing or only a tenuous capacity to claim basic human rights that might challenge how biometrics are applied.

Another key consideration of this book is whether and to what degree security and biometrics are rightly associated. Could a biometric culture actually make many people less safe and less secure? This is a question that goes to the core of how we think about migration and its causes as well as how some kinds of responses to widespread human insecurity maintain threadbare existences rather than alleviate suffering. Even a limited view concerned only with national self-interest cannot help but acknowledge that biometrics may not produce the security it claims. For example, Wendy Brown notes how highly visible security measures in New York City, such as concrete roadblocks installed around notable buildings, worked to unsettle residents after 9/11 (76). Arguably, this is what biometrics does: it creates a culture of fear while establishing a spectacle of security. What kind of ideal of security is experienced as persistent insecurity? If biometrics can count among its effects the regularization of pervasive insecurity, then it is clear that this technology does more than its stated aims of authenticating identity and controlling access.

Is the promise of security enough to make biometrics appealing, especially when we consider that security may never be achieved with more checks, more barriers, more fences, and more aggression toward others? A cynical response might note that some have a vested interest here and that appeal may be a matter of how the technology is sold to the public. There is an industry that stands to profit from fears of insecurity, and governments increasingly normalize apprehension in order to create populations that look to the state for police and military protection rather than look to the state to protect individual and collective rights and freedoms and ensure that everyone can lead a livable life. *Security* may be an imprecise designation of more complex desires, too. The promise of security may be a case in which the answer establishes the terms of the problem, a dynamic Slavoj Žižek has noted by considering how a solution can render social reality more or less legible ("Da Capo" 229). Biometric identity confirmation may teach us that we are surrounded by strangers and that it is necessary to produce knowledge about these strangers and that failing to identify them imperils us in some way. As a solution, biometrics insists that identification makes a difference. Biometrics assures us that the great many people we see on a daily basis are not nameless strangers but are instead individuals with features that can reveal who they are. This promises at least two things at once. It promises knowledge of another despite the apparent anonymity of strangers who make up our collective social life. Second, it promises that we can indeed glean important information about one another by reading them in some fashion. Such non-interaction may be appealing because it allows one to keep others at a distance and participate in their lives at the same time, as if viewing them on a screen. Biometric identification can draw upon a species of destructive attention that reduces individuals to fantasies, coding them as readily identifiable in ways that might recall forms of racism, sexism, homophobia, ageism, speciesism, and ableism that isolate characteristics and produce knowledge about another on the basis of visible markers. Alternatively, if the solution was to greet strangers without demanding they identify themselves first, one might have understood that the problem all along was that we no longer have the opportunity to get to know others and find out why they are here, to share why I am here, and to discover what we hope for individually and collectively. Such a solution would begin from the premise that whoever I am is in some way affected by and affecting who you are. Biometric thought does not trade in scenes of sociability and dependence but instead produces the idea that people are objects to be read and assessed and that anonymity and dislocation can be overcome primarily by procedures of scrutiny that help to invent the very idea that one is a legible object. The appeal of such discrete legibility has proven itself to be quite consistent over at least the past two hundred years, and this promise to read individuals, separate them from ourselves, and penetrate anonymity may go a long way toward explaining our biometric present.

As a mode of organizing reality, biometrics does more than simply verify identity. It regulates and establishes a number of social processes and perceptions regarding how individuals live in relation to the ground beneath their feet. I want to consider two brief examples of biometric identification in order to demonstrate the scope of the arguments of this book. One does not seem to reference biometrics, in the typical sense of the word, and one very clearly involves what we think of when we speak of biometrics. The purpose of these two examples is to highlight some of the long-standing and fundamental questions about social identity and how we exist in a world with others that remain at the core of social forms of biometric knowledge. These examples also sketch out distinct encounters with biometrics to highlight how they can produce quite different results despite their shared ambitions of identification. As a way of framing individuals and the world, biometric thought produces particular dominant and consistent recognitions, but it also opens up other possibilities and presents identification as a process that can operate to multiple ends despite the normative force and institutional aims that often constrict its operations.

Shakespeare's *Hamlet* opens with a provocative series of exchanges as military officers check and re-check identities in the dark of night.

BERNARDO: Who's there?
FRANCISCO: Nay, answer me: stand, and unfold yourself.
BERNARDO: Long live the king!
FRANCISCO: Bernardo?
BERNARDO: He.
FRANCISCO: You come most carefully upon your hour.
BERNARDO: 'Tis now struck twelve; get thee to bed, Francisco.

FRANCISCO: For this relief much thanks: 'tis bitter cold,
    And I am sick at heart.
BERNARDO: Have you had quiet guard?
FRANCISCO: Not a mouse stirring.
BERNARDO: Well, good night.
    If you do meet Horatio and Marcellus,
    The rivals of my watch, bid them make haste.
FRANCISCO: I think I hear them. Stand, ho! Who's there?
Enter HORATIO and MARCELLUS
HORATIO: Friends to this ground.
MARCELLUS: And liegemen to the Dane.
FRANCISCO: Give you good night.
MARCELLUS: O, farewell, honest soldier:
    Who hath relieved you?
FRANCISCO: Bernardo has my place.
    Give you good night.
Exit
MARCELLUS: Holla, Bernardo!
BERNARDO: Say,
    What, is Horatio there?
HORATIO: A piece of him.
BERNARDO: Welcome, Horatio: welcome, good Marcellus.
MARCELLUS: What, has this thing appear'd again to-night?
BERNARDO: I have seen nothing.
MARCELLUS: Horatio says 'tis but our fantasy,
    And will not let belief take hold of him
    Touching this dreaded sight, twice seen of us:
    Therefore I have entreated him along
    With us to watch the minutes of this night;
    That if again this apparition come,
    He may approve our eyes and speak to it. (1.1.1–38)

"Who's there?" is a question that has come to be associated with procedural neutrality. It is the expected question in some form or other when one arrives at a border. It is a question that can present terrific anxiety, provoke ire, seem tedious, or be a prelude to discrimination. Such a simple question backed by a remarkable apparatus of force and technologies of documentation and authentication can produce incredible effects. The question is no longer one that an individual can answer as Bernardo and Francisco once did; it now requires documentation, stored data, and rituals and procedures for recognition and authentication. But it is telling that even in Shakespeare's day, identity was not something full and present in itself. Answering the question alone is not sufficient. It remains a matter of having that answer recognized by another. Shakespeare understood that identification is literally a dialogue; for us, this tends to mean that identification is a social process that involves not just how we choose to present ourselves but also unchosen norms that govern how we are permitted to present ourselves and affirm our identities. These norms include not only the conventions of documented existence, such as one's nationality and likeness, but also more general conventions regarding the social legibility of oneself as a man or a woman or neither. For example, one might not emulate the norms of what it means to appear as a woman, but one is surely affected by unchosen social expectations of gender assignment, even if this might be moderated to greater or lesser degrees depending on any number of factors. Identity depends upon recognition by others, and this means that one will be subject to regimes of recognition supplied by the social world within which identity exists.

If identity is not something absolutely determined by the individual but subject to norms of intelligibility as well as social expectations of appearance, then we ought to follow Shakespeare's lead and note that the environment plays an important role in structuring identity. The context and situations within which individuals are identified matter. In this case, the darkness of night makes identification a more challenging task, and perhaps a more pressing one, too. Biometric identification always involves particular social and historical contexts even if it is routinely premised on dismissing those same contexts, including a natural world and the world of social relations that surround and make possible the life of the individual identified.

Advanced biometric technology promises to identify an individual according to one's physical presence, and this may not be a new phenomenon, as we see in *Hamlet*'s opening lines. While initially a scene of voices calling back and forth, this becomes a moment of identification in which individuals meet face to face. The interaction is set next to the story of a "dreaded sight," an apparition that Horatio does not believe exists but which the guards have witnessed. The first exchange is an account of identification during the changing of the guard, and the second questions what those keeping watch actually see. Perhaps we can read this as a metaphor for the ways in which identification involves a process of recognition that unfolds in two different directions at once. It attaches identity to the body by seeking to discern who someone is by voice and appearance; it also conjures something as if out of thin air—namely, an idea of what a person is and how one can be identified. This idea of a person emerges from the dark night, as if free of attachments and formed in isolation. It is a social concept of identity that biometric identification relies upon as it tries to see who is there. This does not mean that Francisco invents Bernardo, but he is working within a ritual of identification and greeting that is social and which conditions the possibility of this encounter in specific ways. Identification involves seeing that ghostly thing we have deemed a person, but this may not be quite as obvious as it seems. If a person requires particular social and environmental supports in order to survive, is it the case that one's existence is encapsulated by the body? If some can be seen first and foremost as "paperless," itself such a thoroughly strange way to encounter another, is there really a fixed and consistent notion of what it means to apprehend someone? Biometric identification relies upon and affirms particular modes of recognition that conjure some features of identity as visible and important while leaving other features of identity largely unseen and unimportant.

We readily sense, I think, from Bernardo's coy response of "He" to the question of "Bernardo?" that more than one response is possible even to the official demand of identification. Even if biometrics is consequential in its power to regulate identity and constrain access, its power is never absolute and never uninterrupted. It does not fully determine the lives of even those who feel the strongest force of its orbit. Even if biometric identification can demand verification, it does not follow that it can attain what it demands or that its demands are not regularly circumvented in all sorts of ways.

To ask "Who's there?" is to pose a question that the individual cannot possibly answer. Bernardo and Francisco each ask the question of the other and barely answer for themselves. There is little point in doing so. Who they are is for others to record and comprehend. Identity lies outside us, awaiting the recognition of another and the social world of practices and concepts that govern how and whether we are to become legible to others. At the present moment, "Who's there?" can hardly be separated from concerns about terrorism and violence, insecurity and threat, global economics and strategies of impoverishment, migration, and warfare. "Who's there?" has become a question of our age, imbued with new significance while still mediating old concerns about how we produce knowledge and what that knowledge does in the world. It is both a question and an intolerable resistance that must be overcome.

Compare the scene from *Hamlet* to a 2015 media release from the United Nations High Commission for Refugees (UNHCR) regarding the development of biometric identification measures at refugee camps for Burmese citizens in Thailand (Tan). Some of these nine camps have existed for more than thirty years. They were created in response to waves of migration as individuals fled the violence of the military dictatorship across the border in Burma, violence that was especially directed toward ethnic minorities. At these locations, UNHCR has developed the template for enrolling individuals in a system of biometric identification that it expects to make global by 2020:

Between January and May this year, UNHCR and the Royal Thai Government verified and updated the records of nearly 110,000 registered and unregistered refugees from Myanmar with the help of UNHCR's new biometrics identity management system (BIMS).

Thailand was chosen as the first site of the global rollout, complementing plans to verify the bio-data of the camp population. Regular registration in the camps had been suspended ten years ago, and the need to have updated information on family composition, births, deaths, and marriages was growing increasingly urgent.

"The situation in Myanmar is changing and refugees are finding their own solutions outside the camps," said Mireille Girard, UNHCR's Representative in Thailand, noting that small numbers have started to return home on their own. "By understanding their family and individual situation in the camps—including those of the most vulnerable refugees—we can further improve our assessment of their situation before and after any movements. This will also enable us to target assistance and monitor more accurately."

Implemented jointly with Ministry of Interior officials and supported by UNHCR's nongovernmental organization (NGO) partners, the exercise involved close scrutiny of existing documents and physical verification of entire households. Refugee leaders in each camp helped to mobilize the population and encourage them to participate. The result is the most comprehensive protection and statistical review of this refugee population in ten years.

Using BIMS meant that for the first time in UNHCR's history, each refugee's fingerprints and iris scan were collected and securely stored in UNHCR's online database, retrievable from anywhere in the world.

"Biometrics will help refugees in the future as it ensures that once they've been through the system and enrolled with their fingerprints and irises, we'll always know who they are," said Sam Jefferies, UNHCR's Associate Biometrics Deployment Officer in Geneva. "If they lose their documentation, they can always come back to us." . . .

"With these cards we don't need to travel around with heavy equipment like a server," said UNHCR Representative Girard. "In the event of voluntary return, our teams in Myanmar will have a card reader in their backpack when they visit the field to document what has happened to returnees, and if they have received reintegration assistance. We will also pass on that information to other humanitarian actors and the authorities on the ground so that they can plan and deliver services in places where they are needed."

John Smith, a refugee who works for the Karen Refugee Committee in Tham Hin camp in Ratchaburi province, said, "This verification is very important for me and for others who are refugees. The card can be decisive for our life in the future. If we have a chance to go back [to Myanmar], it will be good evidence for us to show to UNHCR or the Thai government. I tell the others to keep their smart card in a safe place, with their most precious things."

Like the opening to *Hamlet*, this is a scene of welcome, though curiously it is designed to welcome UNHCR's new biometrics identity management system more than it welcomes any of the individuals involved. On its website, UNHCR describes this system by noting that "re-establishing and preserving identities is key to ensuring protection and solutions for refugees. By linking new technologies, such as biometrics to existing registration data, UNHCR can strengthen the integrity of existing processes and significantly improve efficiency for operations. Being able to verify identities is extremely important and a matter of human dignity." Biometrics transforms the body into a form of documentation in these camps. If the body is recognized as a means for re-establishing and preserving identity, it is worth considering what it can document and what it cannot. Can a biometric concept of the body acknowledge the conditions that lead one to migrate? Can the idea that the body is a form of documentation account for the ways in which identity is a product of shared relations among family and community? Can it acknowledge that life depends on caring for and being cared for by others? Does recognizing each unique and individual body as the basis for identity help or hinder an understanding of the ways in which identity is never simply of one's own making and thus obscure relations of power and social practices that one does not choose but which are nonetheless profoundly consequential in defining who one is?

The UNHCR account demonstrates that biometrics may produce both valuable and troubling effects. Largely understood as an effort to prevent fraud among claimants of re-integration assistance, this is also an opportunity for individuals to receive official recognition, in some cases for the first time in their lives, and to receive social supports they might not otherwise access. Possessing a simple identity card may not make the water clean or rebuild one's home or guarantee a stable and viable life after repatriation, but it may be a tangible affirmation of one's right to exist without persecution.

Biometric identification is presented as a gift and a means of restoring identity. As a gift—and I do not wish to under-appreciate the supports that official records of existence can make possible—it is a present that may forget what it replaces: namely, a notion of self and community that has been structured and constrained by life in the camps, the experience of dispossession, and a life in Burma before that. This same capacity to forget as a feature of identification is apparent in the report's assumption that biometric identification will lead refugees to return to Burma, as if they could pick up where life left off. For individuals who have grown to adulthood in these camps, the idea of re-establishing one's proper identity and existence back home is remarkably strange. This is not to say that re-integration, restitution, and reparations are not desired and worthy goals; it is simply to quarrel with the biometric ideal that holds verifiable identity to be the only or primary concern in a scenario that involves social, economic, and political upheaval for nations and individuals. To simply guarantee individual identity leaves a great deal more to be done, especially if we are willing to recognize that one does not survive as a body alone. Although biometric technology makes possible varied recognitions that helpfully validate the humanity of an individual, it can also forget the family and cultural histories of existence that go unnamed by the creation of a new official record of one "John Smith," formerly of Burma and now a resident of a refugee camp in Thailand.

Why might one select such a seemingly out-of-place name in these circumstances? Identification, especially a mode of identification that promises to be indelible and written upon the features of one's body, might lead one to conceal—perhaps in order to maintain—the ethnicity that made one subject to persecution. The postwar history of North American immigration is likewise littered with instances of individuals changing their names in order to make their ethnicity less visible. Biometric identity forgets as readily as it records. For some, biometric confirmation might be a welcome opportunity to hide who one was previously identified to be and perhaps also escape the very real physical vulnerability that follows from practices of identification. Identity can exist, under this logic, as a site of tension or memory or something that bears witness to a history of dispossession, perhaps precisely by disavowing that history. What this identity most clearly is *not* is a record of reality as it exists. Not everything can be counted and recorded by the operations of biometric identification; embodied existence is not something purely real that exists outside the larger social conditions that make identity legible.

Biometric thought makes other recognitions, too. What kind of recognition is this that identifies the refugee as a potential criminal and says so little about those who confined them in camps or persecuted them in the first place? Does UNHCR discern in every refugee a potential criminal, a recognition that moves vigorously from the exceptional circumstances of the refugee to imagine that such circumstances are the basis for achieving some kind of unearned gain? It is striking how the threat of fraud aggressively displaces the dispossession and violence that mark the passage of a refugee and eliminates from view the considerable international political and humanitarian failures that almost always subtend mass migrations and the establishment of refugee camps. Examining the particular instance of biometric identification of asylum seekers in the United Kingdom (U.K.) in the early 2000s, Btihaj Ajana notes that such a system provides a public face to human migration and recognizes it to be a matter of identifying "those who are regarded as 'genuinely' potential refugees and those who are perceived as 'bogus'/'irregular' (economic) migrants" (66). While there are important reasons to take fraud seriously, the ways in which criminality and refuge are publicly linked by such operations can overwhelm thought.

There may well be individuals profiting from turning the world upside down, but I have a hard time believing they are refugees. In this example of the Karen refugees in Thailand, the threat of fraud directs money to tech companies rather than to those displaced. Biometric regimes remake the individual according to an economic reality: one exists as someone who receives payments legitimately or illegitimately, not as a person with dignity. Economic concerns receive the priority, and the insistence that the world presently places upon fiscal responsibility trumps whatever else one might have seen in a precarious situation. This emphasis on fraud produces other effects as well. As Katja Lindskov Jacobsen notes more broadly about the growth of biometric technology in this context, "UNHCR trials and pilot projects have given rise to various 'success stories' about the ability of biometrics to advance benevolent objectives," and these "humanitarian success stories can influence the normative framing of biometrics, and thus the normalization and routinization of these technologies" in "policy-making contexts" more generally ("On Humanitarian" 539–540).

At a refugee camp in Thailand, biometric thought creates individuals who can be held to account and made punctually responsible, all the while making unrecognizable much of the unchosen conditions that led to migration, displacement, and the physical and psychic violence that so often attends such experiences. So what begins to emerge from this report is a sense of the varied ways in which biometrics functions and affects how we understand individual lives as well as complex geopolitical realities. It might identify individuals, but it does a great deal more than that, attending to some circumstances while ignoring others.

The biometric registration and identification of refugees in Thailand is immediately different from the opening to *Hamlet* for all sorts of reasons. However, if one focuses on each example as only a scene of identification, some similarities are apparent. Both involve a social process of identification. Whereas Shakespeare created a scene of call and response in which individuals come to recognize one another, UNHCR's operations in the refugee camp are not nearly so reciprocal. Instead, they seem to produce an identity that does not otherwise exist, delivering it to those without official identity. In both examples, relations of power and forces beyond the individual come to define the scenes: because identity is social, it cannot be determined by the individual alone, even if the ultimate aim of such identification measures is to produce an idea of the fully distinct and autonomous individual. In each scene, biometric identification conjures a sense of immediacy and presence, promising to deliver the true identity of the individual. It tries to reduce the question of "who's there" to a certain obviousness that, upon further inspection, it achieves only by insisting that identity is beyond oneself and must be verified, officially, by others. It can be asserted, but it belongs to others to confirm and thus must be presented in an intelligible manner. That is, identity is granted. Such an idea of identity—something that requires others at the same time that it is marked as a property of the body—is a powerful effect of biometric thought. Whether this tension initiates possibilities for thought regarding how one relies on structures beyond one's being in order to be who one is, is something that I am keen to consider.

IDENTIFYING BIOMETRIC THOUGHT

The pages that follow engage a varied archive that I suspect may strike different readers as odd at times. Why discuss fiction in a work on biometrics? Why focus on narrative accounts of individual lives if the subject at hand is technology? Why read philosophical works alongside street art, or histories of face reading next to poetry? My decision to proceed in this interdisciplinary manner reflects my sense that the larger structures by which identity is shaped are never locatable in only one place. Official modes of authorizing identity matter—I could not imagine focusing so much attention on biometrics if I did not hold this to be true!—but so too do cultural expressions of identification, expressions that can police and normalize particular modes of thought and social assumptions, sometimes with as much consequence as official government regulations. I seek to bring these disparate modes of official and unofficial identification into conversation to better understand what is foreclosed or intensified under the ambition of biometric thought. More, to the extent that the social life of biometrics makes the body into a reliable location for identity, it is part of the widespread practice of telling stories that establish the foundations and assumptions that one uses, or which are inflicted upon one, to make sense of reality. It is not the nature of my argument to say that poetry is the answer to the curious forgetting established by biometric practices, like those that identified John Smith in Thailand, as if verse might make losing one's home and culture tolerable. What I do argue is that identification entails a great deal more richness than biometric identification tends to allow. Especially at the present moment—in which identification functions as a strategy that too often rationalizes and perpetuates inequality and injustice—it is vitally important to ask much more than "who goes there" and to offer up much more than oneself in response. My methodological preference for a varied archive asks the reader to be open to the truth that can arrive unexpectedly, like a stranger, in the stories and social norms that can utterly transform us even without having a reliable claim to fact.

The Social Life of Biometrics is an effort to understand the effects of a particular way of recognizing individuals. This might have led me to offer a top-down view of matters that would begin by considering how the biometrics industry understands the technology or how governments apply it. What I discovered is that these official accounts are a great deal less strident than cultural representations of biometric efficacy and that these accounts quickly come to emphasize human rather than technological matters, preferring narrative over science, as becomes apparent in this account from the Biometrics Institute regarding the security of biometrics:

> Hollywood sci-fi movies frequently have scenes where eyeballs are plucked out of skulls, fingers cut off or even whole hands severed to access biometric systems. It's

hardly surprising then that a myth is that a dead biometric is as good as a live one. So how true is this? In most cases the answer is a resounding no, although it's hard to find any volunteers to test empirically prove this. When a biometric is severed, in addition to the trauma inflicted during the removal process, the circulation loss rapidly degrades and deforms the biometric rendering it useless for access control, or for pretty much anything else. In most cases just as with a PIN or password it would be easier (not to mention a lot less messy) to coerce co-operation at gun-point.

The biometric industry tends to publish only moderate claims to efficacy, as this carefully couched answer suggests, complete with luridly distracting details that, if nothing else, associate biometric science with both the image and rhetoric of keeping it real. A 2018 U.K. House of Commons parliamentary report, *Biometric Strategy and Forensic Services*, gave data to support these modest claims when it examined

> the effectiveness of the technology and concerns that its reliability in making matches might vary when applied to people from different racial groups. As we noted in our recent report on Algorithms, research at MIT in the US found that widely used facial-recognition algorithms were biased because they had been "trained" predominantly on images of white faces. The systems examined correctly identified the gender of white men 99% of the time, but the error rate rose for people with darker skin, reaching 35% for black women. In the UK, Big Brother Watch recently reported their survey of police forces, which showed that the Metropolitan Police had a less than 2% accuracy in its automated facial recognition "matches" and that only two people were correctly identified and 102 were incorrectly "matched". The force had made no arrests using the technology.

My sense is that these carefully framed accounts of biometric technology and its actual efficacy in such high-stakes scenarios tend not to fire the public imagination and thus are not the primary sources of biometric thought and the logic of recognition that it establishes. I am not focused primarily on official applications of biometric for other reasons as well. This is not a book that seeks to explain what Muller has called "governing through risk and the emerging biometric state" (200), a phrase that plays upon an apocalyptic rhetoric useful for criticism that seeks to affirm that top-down government-devised operations structure a society. What I offer here is not especially convinced by such suspicions and anxieties and their implied perspective on how power functions. I take a more diffuse view of power that operates in a less top-down coordinated manner and emphasize instead an account of the ways in which culture—including but not limited to the practice of biometric identification—has for some time been working with a biometric logic that attaches identity firmly and often exclusively to the body. Some biometric technology may be new, but many biometric impulses are not; I am interested in thinking about what a tradition of biometric thought does and why it matters and, above all, in highlighting the many different ways individuals have lived under the logic of biometrics.

My examples emphasize a human experience of biometrics. This emphasis is a marked departure from Ajana's *Governing through Biometrics*, in which she contends that biometrics is a technology of governmental control. This thoughtful and engaging account of biometrics is rooted in an idea of biopolitics, a concept that suggests that, since the age of reason and the advent of democratic freedoms, governments have increasing sought to manage freedom by regulating individuals according to social, medical, and political norms that foster life. Focusing on the living being and seeking to maximize life became, for a time, a central means of regulating how individuals live within Western democracies that were defined by their commitments to individual and social welfare. Ajana often emphasizes how practices of biometric identification treat individuals as physical creatures with properties of life that must be managed. Our arguments are closest when she explores how biometric technology can act in excess of narrow applications. For example, on the subject of identity fraud, she notes, "The argument of identity fraud, as such, functions as one of the primary vehicles for facilitating the introduction and spread of identity systems and insuring the public acceptance of them" (114). Our arguments are most distinct when she emphasizes the top-down authority of biometrics. She notes, for example, that the "state's ability to decide on who will and will not be provided with protection turns the domain of asylum into a site of biopower: the power to 'make live' (those who are granted asylum) and 'let die' (those who are deemed undeserving or not genuinely in need of protection)" (60). Such a claim reflects an emphasis on "the fact that the state has a tendency to 'monopolise' the decision on who will and will not be granted protection and determine alone the terms and conditions for such protection" (61). Not everyone falls under this monopoly in the same way or to the same extent, however. I do not hold utopian views about untroubled freedom for those who might cross a border illegally or work without papers, but these are realities that must be counted as part of biometric thought. There are any number of ways that individuals have navigated around governmental aspirations of biometric modes of identity verification, and these suggest that while governments may monopolize practices of official identity verification, they do not control the areas of social life they survey, nor are these the only activities worth perceiving when one casts an eye upon such complex territories. The state may act on its assumptions regarding whose lives matter, but one needs to recognize the resiliency of those who survive despite neglect.

It may be for this reason, then, that Ajana turns to consider storytelling as an alternative to biometric identity formation, noting "that, paradoxically, in its pursuit of capturing the singularity of the person, biometrics only ends up obstructing the exposure of singularity precisely because of its amputation of narrative from the sphere of identity" (106). I agree with much in this claim, though my argument is slightly different. I insist that one never controls the language or social norms in which one speaks one's narrative, and this means that I find untenable an opposition between the individual narrative and an official mode of identity verification, as if one is pure and honest and the other is not. They are simply not opposed, to my mind, but are instead part of a larger process in which identity is always social. This leads me to wonder about the sorts of recognitions any given mode of identification produces. Thus, while a particular feature of this book emphasizes individual interactions with biometric thought, it is not because I am convinced by a humanist recognition of the fundamental truth that emerges from the individual telling one's own story but because every act of identification can reveal the larger social conditions that make individual existence possible.

I will say more about other scholarly accounts of biometrics at various points throughout the book. For now, I will briefly signal that my emphasis on social and cultural expressions of biometric thought is a distinct feature of this work that leads in several directions. First, I insist that the logic and ideas associated with biometrics are shaped by culture, which means that what biometrics might be said to do and how it affects individuals are never limited to its discrete operations. Second, while I chart how biometric culture intersects with discrete biometric practices, I rarely identify government control and regulation as the sole ambition that merits thought and study in such contexts. Third, my focus on a widespread biometric culture necessitates a methodology that emphasizes a wide variety of objects, archives, histories, disciplines, and concerns. This may be most visible in the eclectic selection of materials that structure my analysis. Materials range from Amnesty International reports on migrant workers to eighteenth-century poetry likewise concerned with the economically dispossessed; from pseudoscientific accounts of face reading to the perspectives of bioethicists on facial recognition; and from works of contemporary fiction to street art, each differently engaging the topic of migration. With such diversity, I attempt to document the numerous areas of cultural life that shape and are shaped by biometric thinking on identity.

This book does not offer a comprehensive account of biometrics in culture. Another book might read our biometric present in science fiction and fantasy by charting, for example, the ways in which the television series 24 (2001–2010) did more to cultivate a social foundation for facial recognition as an expression of security than any peer-reviewed research ever could. Could the morbid curiosity of the Biometrics Institute's reflections on severed body parts be explained by the use of detached eyeballs to fool retina scanners in the film *Minority Report* (2002)? What should one make of *Twin Souls*, a project that proposes to match parents with adoptable children in Russia and promises to offer children who are, on the basis of reading the faces of all involved, visually matched to the adopting parents or matched to the face of a deceased child? There is no shortage of ways that one might think about culture and biometrics, so this book cannot claim to be a comprehensive account.

In what follows, I offer a cultural account of biometric technology that seeks to rethink identification by defamiliarizing the attachment between identity and the body that has been such a powerful consequence of the social and cultural life of biometrics in the past two hundred years. The chapters develop these ideas by looking at biometric thought as a concept and a set of effects associated with its operations. What are its domains and areas of influence? What kind of recognitions does it make possible? What is its history and how has it changed over time? Are there alternatives to biometric thought that might differently apprehend a similar conceptual territory?

The first chapter introduces a series of accounts of what it means to live a life directly affected by biometrics. Each individual's story helps to chart the reach of biometrics and how it can regulate so many different human experiences. These narratives document and theorize the force and impact of biometrics upon an individual as well as the ways in which biometrics makes and remakes identities and relationships.

Chapter 2 develops the social life of biometrics. *Biometrics* refers not only to a set of practices of identification but to a much wider sense of meanings and actions in the world, such that it can be said to be a mode of thought. The chapter insists that what is needed at present is the capacity to ask not just "What is biometrics?" but "What does biometrics do?" To document some of the effects of biometric thought, it looks at media accounts of biometric technology and migration as well as scholarly accounts of surveillance. Another topic the chapter addresses is the idea that biometric thought is a particular challenge to those who would insist that biometrics is a technical and therefore neutral means of establishing and authenticating identity.

The idea of technological neutrality is only one domain central to how we understand biometrics, and chapter 3 examines a number of other conceptual attachments that the social life of biometrics intensifies and directs anew. What particular social concerns tend to be addressed by biometrics, and how does biometric thought constrain what these areas of concern mean? According to its influence, what becomes obvious and what disappears from view? After identifying how concepts such as security, transparency, humanism, presence, borders, and identity come to be central to biometric thought and its operations, I consider two particular instances. The first is a Canadian law that would unveil Muslim women at citizenship ceremonies. The second is Yuri Herrera's novel regarding migrants crossing the Mexico-U.S. border, translated into English as *Signs Preceding the End of the World*.

The fourth chapter develops and explicates the methodology that I develop over the opening chapters by charting what is involved in approaching biometrics as a social practice of recognition. I realize it is unusual to withhold a chapter on method until halfway through a book. My sense is that the earlier chapters need to develop key ideas first, after which chapter 4 will establish and enunciate the principles that structure the book as a whole by asking, What are some of the most powerful recognitions of biometrics, and what do these do in the world? How are lives apprehended by biometric thought? At a time when refugees are identified as "human garbage" by Janusz Korwin-Mikke, a Polish member of the European Parliament, this chapter considers how biometric thought regulates what we can recognize about human movement, migration, and identification (BBC News). I consider the difficult instance of Nilüfer Demir's photograph of three-year-old Alan Kurdi, who was found deceased on the coast of Turkey in 2015. In what ways were the meanings of this image regulated by biometric thought such that it made some matters powerfully recognizable to the world and others more difficult to see? I close the chapter by distinguishing this method from existing critical accounts of biometrics.

In chapter 5, I offer a brief history of the social life of biometrics, tracking the development of passports and fingerprinting, and I place special emphasis on the emergence of a European culture of physiognomy, or face reading, at the end of the Enlightenment. This chapter notes that biometric thought draws upon shifting social and political priorities that shape how it regulates identification and human mobility. To illustrate this shifting history, I focus on three instances of biometric thought. One confronts the recent controversy regarding the national identification of Olaudah Equiano, the author of the first English-language slave narrative, in order to consider how this betrays biometric ambitions then and now. The other two are poems by William Blake and Percy Shelley that highlight how individuals have long contested biometric norms of recognition, and each offer alternative ways of seeing individuals.

The final chapter considers what it means to think in the wake of biometric thought and wonders what sort of future is possible despite and because of its influence. Beyond the rehearsed choreographies of identification, what else might biometrics make visible? Thinking with and against Giorgio Agamben's reflections on biometrics, I ask: What kind of ethics and responsibility are possible under biometric thought? This chapter entertains the prospect of thinking in ways that contest biometric truths regarding human identity, migration, strangers, borders, and the nature of the body. Of particular interest here is a philosophy of hospitality that might offer a new set of possibilities distinct from the dreams of biometric thought while still confronting questions of human mobility.

Biometric thought searches out the conditions within which identity will be authenticated, including the social norms, practical measures, and regulatory concepts that will support that operation. It dreams of a passport photograph made real, in which individuals exist as if floating free against a blank background, each one utterly singular, each one precisely comparable to the other, each one static. It is a dream that we may not be able to halt, because we can scarcely imagine what it would mean to wake up in a world without it.

1 · BIOMETRIC ENCOUNTERS

A commonsense concern with biometric technology is that it is invasive and violates a person's right to privacy or that it might do so if data is stored improperly or stolen. I have little quarrel with this legitimate concern except to say that this should not be the primary or only concern one has with biometrics. The idea that biometrics may compromise an individual actually conceals a very different insight: many of the most powerful effects of biometric measures are social rather individual. As a practice of verifying identity, biometrics depends upon a set of assumptions that frame and produce a given understanding of reality, and these norms do not belong to or issue from any single individual. I refer to *biometric thought*, then, as a way to address all of the ways in which biometrics installs and reinforces a set of norms that regulate how identity is attached to the body and how human movement is controlled and understood.

Biometric thought sounds, perhaps, like the name given to a set of reflections upon the practice of biometrics. It suggests that there are activities that involve biometric technology and then that there are ways of thinking shaped by those practices. I want to challenge that linear order and suggest instead that biometric thought is what makes those activities possible, even as this way of organizing and representing reality is in turn shaped by those practices. Key to this perspective is the idea that biometric thought is not a settled and fixed manner of representing reality but an ongoing way of mediating social life and ensuring it makes sense in particular ways. Biometric thought is what makes possible the ways in which one interacts with biometrics and how biometrics interacts with a given population, shaping how one thinks about security, national borders, the physical borders of bodies, the movement of people across the globe, identity as something that is definitively one's own, surveillance and observation, and how one documents oneself and others. This is not to say that biometric thought is the only social discourse that affects how one encounters these matters, but it has become what Michel Foucault once identified in another context as "an especially dense transfer point in relations of power" (*History* 103). Biometric thought collects and weaves together a number of distinct concepts and areas of social life and intensifies them by directing them toward operations of biometric authentication and its associated power to regulate identity.

In order to contemplate this broader understanding of biometrics, this chapter presents a series of human encounters with biometric technology and biometric thought. The instances I discuss here capture a range of biometric technologies, from passports to face reading and from DNA to fingerprint scanning. The premise underlying this chapter is that biometrics is not a single force that affects everyone in the same way. Because it draws upon and mediates a wide range of knowledge and regulatory ambitions regarding how individuals exist and what one can know about oneself and others, biometric thought must be situated in the particular instances in which individuals encounter or navigate some sort of expectation of identification, sometimes supported by the force of a law but also often supported by normative social expectations. To the extent that biometric thought conditions a social concept of identity and makes it legible as such, there is always more at stake in biometric procedures than the direct process of confirmation that technical manuals of biometrics refer to as one-to-one matching (confirming the identity asserted by an individual) or one to many (identifying a present stranger). Biometric thought is rooted in such acts but it also produces and reflects a number of social presumptions, reinforcing them or presenting them as if they are natural and commonsense conclusions while ignoring alternatives that are made less imaginable by its operations. What does it mean to live in the midst of such circumstances?

SID HILL

"I'm a citizen of the six nation Haudenosaunee, commonly known as the Iroquois confederacy, one of the original peoples of what is called North and South America. We travel the world on our own passports, embracing the full rights extended by the rules of international law and diplomacy. Too often, our passports are denied by the very countries that took our land. They call them 'fantasy documents,' but they are not" (Hill). So writes Sid Hill, the Tadodaho, or traditional leader, of the Onondaga Nation, in *The Guardian* in October 2015. He recounts an instance in which he was prevented from traveling home from the World People's Conference on Climate Change and Defense of Life in La Paz, Bolivia, because the plane would transit through Lima, and Peru refused to recognize his Haudenosaunee passport, possibly at the behest of the Canadian government:

> This is not the first time we have encountered such problems. The best-known incident happened in 2010, when Great Britain refused to allow the Iroquois Nationals lacrosse team to travel to the World Lacrosse Championships in Manchester on Haudenosaunee passports.

Back then, when the British asked the Americans if they would let us return after the tournament, the Americans were in a bind. Saying yes would officially recognize our passports. And it was absurd to say no, since this issue began when the Europeans arrived five centuries ago and seized our lands in the first place.

So the Americans said nothing, leading to a days-long standoff before then-secretary of state Hillary Clinton, a former New York senator, granted a one-time waiver. That was still not enough for the British, so we were barred from the championship of a game we invented and which is central to our culture. (Hill)

Hill's account highlights a great deal about the work of passports, one of the most commonplace devices of biometric identification. His experience suggests that for him, traveling on a passport and confirming his identity is not a simple procedure but one that involves forms of historical memory and conflicts over who can recognize the legitimacy of documented existence. When nations refuse to recognize Haudenosaunee passports, it reiterates the dispossession of the Onondaga Nation. *Identification*, in this context, is the name given to a relation of power and the legacies of colonialism that have invalidated the authority of the Haudenosaunee to govern themselves as a nation and a people. If biometric identification involves verifying official data, who has the power to create that data and what sorts of records are deemed legitimate matters a great deal. The effects of biometric inspection are not limited to comparing pictures to faces and checking travel itineraries. It involves histories of conflict and dispossession as well as conflicts over sovereignty, and its operations may perpetuate colonialism by continuing to forget that history.

What it means to verify and record information associated with identity can entail a great deal more than inspecting a person, then, in both its aims and its effects. At borders, identification may involve not seeing or refusing to see as much as anything else. As Hill notes, these measures constitute a larger refusal to see and respect the Haudenosaunee as a people. The immediate disrespect the Haudenosaunee face at the border is a symptom of a larger refusal to acknowledge their legitimacy as a people. The lived "inconveniences" of the failure to recognize national passports are minor, he notes, "compared to centuries of struggle to maintain our standing among nations of the world."

This scenario also reminds us that passports can have important nation-building effects. Hill writes,

Maintaining our sovereignty demands that we use our own passport. This is why we stamped the passports of visiting nations—including US Americans and the British—in September when the World Indoor Lacrosse Championships was held for the first time on Haudenosaunee land: to underscore that this has always been and remains our land.

We do not have the option of simply accepting American or Canadian passports. We are citizens of the Haudenosaunee Confederacy, as we have been for mil-
lennia before the Europeans' arrival.

That is not negotiable. (Hill)

If biometric procedures tend to isolate individuals and treat them as mobile units
capable of moving about the globe, Hill finds that this perception depends upon
a mistaken assumption that territory and sovereignty always align perfectly. Before
individuals are ever represented by their faces in passports, biometric thought
works to erase the land and insist that each one of us exists equally and can in the
same way be represented as citizens of a nation. Losing this context means violently forgetting the history of colonialism and the Haudenosaunee Confederacy's existence as a nation. Asserting the right to document its citizens, as Hill does,
can mean challenging such forces of erasure by refusing to ground identity in a
deliberate act of forgetting.

HARUN AL-RASHID

Biometric thought promises to reveal. It confirms identity and establishes trust
on that basis. Concerns surrounding national security have cemented assumptions
that transparency and openness are the foundations of moral honesty and civic
life, particularly in the West, just as they establish that concealment and dissimu-
lation are causes for concern. Curiously, this view is contradicted regularly within
popular culture: unnoticed random acts of kindness, hactivists such as Anonymous, and fantasies about masked superheroes whose social value is indexed to
their anonymity. There is a powerful cultural undertow drawing attention to
the possibility that if openness is desirable, it does not follow that the opposite is
true. This has long been the case. The Bible tells of numerous figures who assume
disguises for instructive ends, including the well-known call to hospitality: "Be
not forgetful to entertain strangers: for thereby some have entertained angels
unawares" (Heb. 13.2).

Harun al-Rashid, the eighth-century caliph of Baghdad who is known to
English readers as a hero of Richard Burton's Victorian-era translation of *One
Thousand and One Nights*, walked incognito as a beggar among his subjects. Like
many historical figures who have entered popular culture, this version of Harun
likely bears little resemblance to the actual person. But that in itself is instructive.
What is it that is so appealing in the figure of a leader who mixes unseen among
his people? As every politician knows, such representations humanize individuals. And what a curious admission this is: that being human is the effect of per-
formances that in some way humanize oneself, which is to say one performs a
norm of what it means to be human. That one must act in a way that appears
human might sound odd, but there is a long history of doing so. For Harun, just

as it is for so many others, becoming human is a process of putting on a mask that can be recognized by others. Biometric culture tends to evoke the opposite: we must strip back all the layers of how one might present oneself to arrive at an essence. But perhaps the very idea of an essence is yet another mask, especially if it is one that relies on others to discover the truth beneath the surface.

In the novel *If on a Winter's Night a Traveler* (1981), Italo Calvino writes a provocative opening, and nothing more, to a story that might have taken place alongside *One Thousand and One Nights*:

> The Caliph Harun-al-Rashid . . . one night, in the grip of insomnia, disguises himself as a merchant and goes out into the streets of Baghdad. A boat carries him along the waters of the Tigris to the gate of a garden. At the edge of a pool a maiden beautiful as the moon is singing, accompanying herself on the lute. A slave girl admits Harun to the palace and makes him put on a saffron-colored cloak. The maiden who was singing in the garden is seated on a silver chair. On cushions around her are seated seven men wrapped in saffron-colored cloaks. "Only you were missing," the maiden says, "you are late"; and she invites him to sit on a cushion at her side. "Noble sirs, you have sworn to obey me blindly, and now the moment has come to put you to the test." And from around her throat the maiden takes a pearl necklace. "This necklace has seven white pearls and one black pearl. Now I will break its string and drop the pearls into an onyx cup. He who draws, by lot, the black pearl must kill the Caliph Harun-al-Rashid and bring me his head. As a reward I will give myself to him. But if he should refuse to kill the Caliph, he will be killed by the other seven, who will repeat the drawing of lots for the black pearl." With a shudder Harun-al-Rashid opens his hand, sees the black pearl, and speaks to the maiden. "I will obey the command of fate and yours, on condition that you will tell me what offense of the Caliph has provoked your hatred," he asks, anxious to hear the story. (257–258)

Told with the logic of a dream in which individuals and structures seem to unfold before Harun as he moves through the scene, this is a tale about stories and how they structure a life. Harun does not know what offense he has committed. Assuming he is not simply inattentive and unaware of how his reign affects his subjects, this points to the possibility that Harun's identity is not so firmly attached to his body, such that some attribute to him actions he would not claim or understand himself to be responsible for. This same insight is repeated by the misrecognition that leads Harun to be invited to sit with this group in the first place. In neither instance is his knowledge of himself authoritative or sufficient. The narrative demonstrates the extent to which identity is outside of oneself, in the hands of others, and beyond one's control.

It also describes the sense that official identity may emerge out of narrative conventions or what Groebner calls the "aspects of fiction . . . inherent in systems that endeavored . . . to record real historical beings on paper" (219). The actual Harun al-Rashid likely did none of the things attributed to his many fictional portraits, and this mode of telling a story may not be so different from what official records of identity do when they select some details as relevant in order to track who one really is in the world. And it is on this basis that the threat of the black pearl is so disturbing: knowing that others will act on the basis of a fiction and possibly forget that it is an invented or partial account of who one is.

Disguise can mean a great many things. It has come to be seen as that which biometrics alone can overcome. But biometric identification is a way of telling stories about a person that, however true they might be, may not align with other modes of identification. It may veil as much as it unveils. It may even function as a black pearl that instructs others to act on the limited knowledge it records. Perhaps what the example of Harun demonstrates best is that disguise is always a question of who has the power to identify what is and is not a disguise. What openness means and what concealment suggests are far from settled matters, but discussions of biometrics frequently make it seem as if they are

ELIZABETH CRAVEN

In 1789, Lady Elizabeth Craven published her travel narrative, *A Journey through the Crimea to Constantinople*. Presented as a series of letters, it records her travels eastward beginning in France and culminating in the Ottoman Empire. A recently divorced aristocrat, Craven was a notorious celebrity, and her narrative celebrates her newly found independence from her husband, from whom she separated after both had very public affairs. Much of Craven's account of Turkey at the end of the eighteenth century involves forms of sociality mediated by the intercession of the veil, an object of curiosity, desire, and intense fascination in the book. The reasons for this are several, not the least of which is the promise of freedom that such an article of clothing might provide an English woman whose movements had become the subject of public gossip.

It might seem strange to reach so far into the past in a book concerned with biometrics, but part of my argument is that biometrics draws upon existing cultural assumptions about visibility and transparency, alongside how such concepts mediate identity. In short, biometrics has a history, and Craven's interest in veiled faces reveals some of the complicated assumptions that lie at the core of this history.

Animated by what Filiz Turhan identifies generally as the "desire and disdain" common to British representations of this region (41), Craven's narrative is an early expression of preoccupations with the East that coincide with Britain establishing its empire in the region. Writing about the Orient was largely a mode of hallucinating Middle Eastern and Indian cultures, and eventually Pacific Rim cultures, for the British imagination. I say *hallucinating* because such writing did not describe the Orient as it was, but instead offered what Edward Said has called Orientalism, a mode of seeing what one chooses to see there. While the Orient and the Occident are academic terms for an empirical reality, Orientalism refers to the widespread social and cultural project that invented a reality it claimed to represent. It was a way of speaking about the Orient, "authorizing views of it, describing it, by teaching it, settling it, ruling over it: in short, Orientalism [was] a Western style for dominating, restructuring, and having authority over the Orient" (Said 3). My attention to biometric thought is inspired by Said's recognition of the ways in which regimes of truth that frame reality are the product of a range of social modes of knowing and ways of acting on that knowledge.

Craven exemplifies early expressions of prejudicial recognitions of the nature of life in the Middle East that have not yet ended in the West. Thus, it is not surprising that upon visiting the Hagia Sophia mosque, Craven offers few kind words for what was once a Greek Orthodox church and also briefly a Roman Catholic cathedral:

> The dome of St. Sophia is extremely large, and well worth seeing, but some of the finest pillars are set topsy turvy, or have capitals of Turkish architecture. In these holy temples neither the beautiful statues belonging to pagan times, nor the costly ornaments of modern Rome, are to be seen: some shabby lamps, hung irregularly, are the only expense the Mahometans permit themselves as a proof of their respect for the Deity or his Prophet. I went and sat some time upstairs, to look down into the body of the temple—I saw several Turks and women kneeling, and seemingly praying with great devotion. Mosques are constantly open; and I could not help reflecting that their mode of worship is extremely convenient for the carrying on a plot of any sort. A figure, wrapped up like a mummy, can easily kneel down by another without being suspected, and mutter in a whisper any sort of thing. (286)

With startling efficiency, Craven moves from the spectacular sight of what is today a museum in Istanbul to the threat of Turkish plots. She anticipates a form of biometric thought that identifies Muslims as potential terrorists first and individuals with human dignity a distant second. I don't share this dispiriting stereotype to note just how pervasive it is (we require little further evidence of the West's strangely insatiable appetite for Eastern terrorists) nor how old such habits of thought are. What I find worth noting here are the conditions that govern this perception. The individuals she sees are threats precisely because they cannot be apprehended properly, which is to say, cannot be apprehended according to Western norms of visible, open faces. Craven fast-forwards from covered heads and draped veils to the specter of visible invisibility, showing the work of unreason that persistently controls such discussions. Why should unseen faces provoke concern at all? Craven could scarcely have expected to recognize the strangers she encountered on her travels.

Craven begins to see another possibility in the veil and unexpectedly discovers the comparative freedom the veil could offer an English noblewoman accustomed to having her movement monitored and gossiped about. Like Lady Mary Wortley Montagu before her, who also published an account of her travels in the region, Craven finds herself thinking about the ways her public life is policed by social expectations of normative aristocratic femininity and sees the practical advantage that comes with being unidentifiable. She explains that in Turkey, "women are perfectly safe from an idle, curious, impertinent public, and what is called the *world* can never disturb the ease and quiet of a Turkish wife—Her talents, her beauty, her happiness, or misery, are equally concealed from malicious observers" (304). Writing in 2008, Maliha Masood comments similarly, "The veil denies men their usual privilege of discerning whomever they desire. By default, the women are in command. The female scrutinizes the male. Her gaze from behind the anonymity of her face veil or niqab is a kind of surveillance that casts her in the dominant position" (226). These perspectives may overemphasize the outcomes of contesting conventions of female visibility and display, but I agree that there is a powerful reversal at work here, even if the reliable consequences of veiling may be uncertain.

Craven's varied reflections on veiling express a mode of biometric thought that captures some of the complexity of what it means to look at a face. Arguably, the face is one of the oldest biometric identifiers and among the most powerful ways of representing individuality in the West. Craven affirms a notion of moral openness by suggesting that covering the face is always a form of disguise—whether dangerous or liberating—yet the prospect of eluding the eyes of others is nonetheless powerfully appealing. What draws her to the veil in either instance, perhaps, are the ways in which power flows through channels carved by ideas of social visibility and the possibility that the social practices of looking and being looked at can be manipulated. Biometric thought might prefer to imagine scenes of inspection in which it is clear who has the power to assess and inspect and who does not, but Craven's reflections long ago established the ways in which these relations of power can shift and change, and this remains a feature of biometric identification even now.

Craven's fascination with veiling highlights another important feature of biometric thought: it rarely operates alone. In this case, physical dress, gendered social codes of public propriety, and the morality of transparency all combine to form a social field of identification in which veiling becomes particularly provocative. This also means that biometrics involves more than discrete acts of identification; it involves a culture concerned with how individuals appear and the means by which their visible identity is regulated, such that one can appear to be a presumptuous foreigner dressing like the locals, or appear to be a victim of oppressive codes of dress, or appear sexualized or de-sexualized by making oneself less visible. And all of this can be the effect of a simple piece of cloth.

The veil is a contentious fabric within the West—at least as much now as it was then—and as such, it mediates a number of concerns, not the least of which is the idea that fabric can mark a decisive and fundamental separation between Eastern and Western cultures, marking one as regressive and the other as liberal. Craven wondered if a veil could liberate women from unwanted attention by enabling them to move in public unobserved. Taking this position is more difficult to imagine today when the veil is frequently represented as an expression of patriarchal control over women. Perhaps such a view is heard so often, in part, because it helps to disavow the entrenched patriarchy of Western cultures. One need only look at rates of pay for women or the abysmally low rates of conviction for rape to capture how intensely normal it is to discriminate against women in the West. Thus, when the veil is held up as proof of patriarchal discrimination—proof, moreover, that such discrimination is the property of another culture, not this one—one has to sense a certain effort at displacement that encourages a failure to recognize what a society designed to value men ahead of all others looks like.

If biometric logic looks to the body as the ground for identity, Craven shows that what it means to be seen is far from universal. As Craven well knew, to be a woman in public was to be subject to scrutiny and to receive unsolicited attention from others directly and indirectly. The very prospect of visibility has a history, then, that includes social practices and assumptions regarding how one's appearance is regulated and what it can be said to indicate. Could biometric ambitions to see and read individuals have their origin in gendered practices of regulating especially the movement of women? It would be a difficult claim to prove with certainty, but it is a possibility that helps to keep in view how biometric operations of attaching identity to the body depend upon supports and social norms that reflect the broader operations of a culture. Biometrics does not operate in a vacuum. Its fantasy of natural visibility and openness encodes a very particular understanding of what the face is, what it does, and the conditions that govern how and in what ways it appears. To present oneself for inspection may be the most powerful form of veiling ever imagined.

KIESE LAYMON

In a well-publicized piece published on Gawker.com, "My Vassar College Faculty ID Makes Everything OK," English professor Kiese Laymon did at least two strange things. He confessed his ambivalence toward the protections that his faculty ID card provided him from some of the racism faced by African-American men, and he confessed on behalf on that ID, recognizing that it has a life of its own of sorts.

While his ID could not prevent him from experiencing the hypervulnerability African Americans experience in the United States thanks to a culture of white supremacy that sees black lives as disposable—when it sees them at all—it did ensure his life was different from those who live without such protections. Considered as a powerful example of how identification operates within a particular context, Laymon offers insights that are key for thinking about biometric thought:

> The fourth time a Poughkeepsie police officer told me that my Vassar College Faculty ID could make everything OK was three years ago. I was driving down Wilbur Avenue. When the white police officer, whose head was way too small for his neck, asked if my truck was stolen, I laughed, said no, and shamefully showed him my license and my ID, just like Lanre Akinsiku. The ID, which ensures that I can spend the rest of my life in a lush state park with fat fearless squirrels, surrounded by enlightened white folks who love talking about Jon Stewart, Obama, and civility, has been washed so many times it doesn't lie flat.
>
> After taking my license and ID back to his car, the police officer came to me with a ticket and two lessons. "Looks like you got a good thing going on over there at Vassar College," he said. "You don't wanna . . . ruin it by rolling through stop signs, do you?"
>
> I sucked my teeth, shook my head, kept my right hand visibly on my right thigh, rolled my window up, and headed back to campus.
>
> One more ticket.
>
> Two more condescending lessons from a lame armed with white racial supremacy, anti-blackness, a gun, and a badge. But at least I didn't get arrested.
>
> Or shot eight times.
>
> My Vassar College Faculty ID made everything okay. (Laymon)

Can we consider Laymon's account of his faculty ID as a narrative of the power of biometric thought? It is clearly a narrative of survival in the face of endless racism and an acknowledgment of the emotional scars that are borne so unevenly within a nation that knows it "can't structurally and emotionally assault black children and think they're going to turn out OK" (Laymon). Reading this as an account concerned with identification does not mean ignoring those matters, but it may mean apprehending them differently, and this immediately reveals how biometric thought organizes the reality it seeks to record.

Laymon recognizes that his life is significantly altered by this thin piece of plastic bearing his image and the seal of Vassar College. It is an object that does more than identify him. His driver's license or passport could do that, but neither provide the safety that his faculty ID affords, which is why he presents it along with his license. So, not all identification is created equal, and what it can accomplish varies. Identification alone does not capture all the potential work an ID can do and what it can say about the person. An ID can create or alter the world within which one lives. In this instance, it has the power to regulate how Laymon appears to others and thus alters his experiences and what it means to survive in a nation in which white supremacy has been normalized in exceptional and ordinary ways.

If we only understand biometrics to be a way of authenticating identity, we will fail to notice how it produces what it claims to represent. Worse, if biometric thought is allowed to erase racism from consideration by affirming that all it does is match an individual to a legal identity, what kind of violent denial of the conditions that shape history and the present does this constitute? Identification means being exposed to norms of social identification and legibility and who one can be seen to be. In Laymon's case, his faculty ID ensures that he can be seen to be other than what representatives of a racist society are willing to allow him to be as an African American. Biometric identification is not simply technical. It has a social life, and it supports and is supported by existing relations of power. More than a means of documenting identity, an ID can document and transform the ways in which one experiences identity, if not recording then at least acknowledging some of the relations of power that shape how one lives.

Laymon notes that the difference between having this ID and not having it can be the difference between jail and freedom or even life and death. If identity is always, in part, a social practice of recognition, it means that the ID has the capacity to produce a new identity for Laymon. It makes him respectable in the eyes of police who might otherwise find him expendable. Identification does not simply record identity; it transforms it according to the power and privilege that can be attached to it or rendered conceivable in its wake.

Laymon's essay is not about identification but about the ways in which we live beyond, in spite of, and because of our identities, all at the same time. This is not a universal experience, as if we are all equal before the white backdrop of photo IDs. Laymon reveals that the experience of being subject to identification and its revelations is far from neutral and depends on the social norms and histories of racism of a given society. What my faculty ID does for me as a white male in Kelowna, British Columbia, is not nearly as dramatic, for example, because I already appear to embody what my society considers to be normal. I never need to counter someone's assumption that I could not possibly be a university professor. What an ID does is in excess of what the ID can record and make visible about a person. In a sense, an ID takes on a social life of its own and does things in the world that are entirely separate from the person it claims to represent. All of this is to say that under biometric thought, it is necessary to part with the idea that identification is merely a representation of a person; much more than this collects around and circulates through biometric technologies of identification.

What kind of cultural memory does an ID capture? Laymon notes his path has been guided by a rich tradition of African-American intellectuals as well as by his family. To his story, he adds the narratives of so many others who experience the racism that his ID protects him from at present. This social context can be addressed only because he included it. It is not included in his ID. At most, his

ID can include the sort of normalizing, vicious context offered by the racist police officer who recognizes him to be exceptional. The ID might protect him, but it cannot give an account of his life. It cannot remember the pain of his mother's generation and how that affects a child growing up.

Laymon's essay is about more than identification, then, in the sense that it refuses the isolating power of biometric thought that seeks to excise individuals from networks of experience and support as well as from environments of harm that shape their lives. His approach insists that identity is always experienced with others and according to social norms of recognition that are ready to discriminate in subtle and overt ways. When the racism that Laymon and many others faced at Vassar was denied by university administrators, he learned that was so normal as to be unrecognizable for those who could not feel its force:

> I expected that four teenage black boys from Poughkeepsie would have security called on them for making too much noise in the library one Sunday afternoon. I expected security to call Poughkeepsie police on these 15 and 16-year-olds when a few of them couldn't produce an ID. I expected police to drive on the lawn in front of the library, making a spectacle of these black boys' perceived guilt.
>
> A few days after Vassar called police on those children, a police officer visited one of the boys while he was in class and questioned him about some stolen cell phones and iPods at Vassar. When the kid said he didn't know anything about any stolen cell phones, the officer told the 15-year-old black child, who might have applied to Vassar in three years, to never go back to Vassar College again.
>
> I didn't expect that.
>
> Vassar College, the place that issues my faculty ID, a place so committed to access and what they call economic diversity, did its part to ensure that a black Poughkeepsie child, charged with nothing, would forever be a part of the justice system for walking through a library without an ID.
>
> There is no way on earth that a 15-year-old child visited by police officers at his school for walking through a local college library while black is going to be OK. (Laymon)

As if this were not maddening enough, he notes that the privileges of his ID did not liberate him from a world of arbitrary arrest and violence but intensified his fear of such senseless violence. He knew that his "family needed me home. My soul needed to be there. But I was afraid to be somewhere where my Vassar College Faculty ID didn't matter worth a damn" (Laymon).

What can an ID do? What does it make possible, and what does it make less possible? To pose these questions, and to answer them as Laymon does, is to show how much biometric thought leaves out when it acknowledges only the individual choreographies of inspection and authentication that characterize identification, which it assumes to be seamless, consistent, universal, and reliable. If biometrics imagines that it takes hold of the body, Laymon suggests that we have barely begun to grasp the complexity of what is involved in actually seeing what embodied existence entails.

Biometric identification establishes the unique and individual body as the ground for identity. Yet for me and Laymon, how we are read and whether or not our bodies can signify *university professor* is not an individual matter but a matter of social categories of race, class, gender, and age and the power and privilege that are attached to each. Identification of the unique individual is clearly not the same as identity and its reliance on social categories of existence. Yet it makes little sense to speak about them as entirely separate, either.

SURIYA

Suriya traveled from Nepal to Qatar as a metal worker employed to build Khalifa Stadium, site of the 2022 World Cup. Amnesty International interviewed him in 2015, along with many other migrants who arrived from South Asia. "When I arrived in Qatar one of the men from the company met us at the airport and took us to the camp. As we departed the airport and went to the company bus in the parking lot they took our passports. Nothing was explained to us, he just demanded our passports" ("The Ugly Side" 23). While Qatari law and the Supreme Committee's Workers' Welfare Standards "prohibit employers from retaining migrant workers' passports," Amnesty notes, "the practice is commonplace" because it enables "employers to have significant influence over their workforce, for example, by threatening not to return them and preventing workers leaving Qatar" (23). Withholding passports is just one tool for maintaining absolute control over migrant workers. Employers also routinely fail to renew work visas, which has the effect of canceling a worker's visa ID. Under Qatari law, such a worker becomes illegitimate and can be arrested and detained as well as charged a fee of more than $13,000 (USD) in order to leave the country under such paperless circumstances (34). Facing an exorbitant fee on exiting the country or unable to access their own passports, migrants are effectively denied mobility and the ability to leave if they are unwilling to work for their less-than-promised wages or in roles other than those for which they were hired. These rules became especially difficult in April 2015 when a devastating earthquake struck Nepal. Migrants were unable to leave in order to care for—or search for—family and friends.

A United Nations report on migrants in Qatar, presented in 2014, noted similar practices among workers employed in domestic labor: "the vast majority had had their passports confiscated" (*Report* 11). Women are forced to endure sexual abuse as well as work as many as 21 hours per day with little recourse. Without access to documentation, migrants such as Suriya are effectively stuck, unable to leave Qatar and thus unable to use their only leverage to demand safe working conditions and fair and timely compensation. Neither biometric inspection nor biometric thought is the direct cause of what amounts to forced labor. But this scenario reminds us that proper documented identity can be a means of constraining freedom as readily as it enables it. To speak of the documented versus the undocumented, for example, makes little sense if this is meant to capture an absolute difference in one's experience of human mobility. Biometric identification operates in this context as a mechanism for regulating movement, not simply as a technique of permitting or prohibiting it. It demonstrates, as well, that identification depends upon contingent laws and relations of power. It is not a form of power in and of itself if those who have legal documentation are nonetheless prevented from being able to move from one country to another.

OMER KIYANI

Biometric identity verification has a range of potential applications beyond passports and border controls, including regulating who can access computers, offices, phones, and even guns. This was what Omer Kiyani saw when he engineered and developed a biometric trigger lock for handguns in 2014. In response to recent and regular mass shootings, the United States government announced new funding in 2016 for research into biometric gun controls. Kiyani's device requires the user to provide a fingerprint to unlock the weapon. Others have created prototypes that involve applying the unique grip pressure of an authorized user in order to make the gun functional.

What is so remarkable about these propositions is how they turn to technology to solve social problems and a long-standing culture of violence. While often presented with a spirit of ingenuity and even awe, this faith in technology amounts to a celebratory refusal to engage with social problems that lead individuals to kill one another (and citizens of the United States do so at wildly disproportionate rates compared to other parts of the world).

Biometric locks do not alter a culture of gun violence and do not look toward a world in which it is not lethally easy to destroy others and oneself. By addressing symptoms alone, such technological fixes ensure that the underlying conditions remain unchanged; thus, in this instance, biometrics promises to make enduringly ruinous conditions more sustainable. A regularized culture of death ought to demand any and all answers without delay, and if practical technological fixes can be readily implemented, they should be. But that should not be all a society does, either. Technology should not come to replace the pursuit of a just world.

As an example of biometric thought, Kiyani's device shows that biometric practices need to be understood not just for what they directly accomplish but also for what they make less conceivable, such as a future in which a nation is not armed for domestic warfare but instead spends as much time, effort, and money on securing the conditions of peace.

ANSELIM UDEZUE

Groebner records the remarkable story of Anselim Udezue, who acquired visas and traveled across Germany, Austria, and Switzerland in 1998. It was only upon his departure from Zurich that someone noticed that his otherwise legitimate-looking passport was issued by British Honduras, a country that does not exist. As remarkable as this fraud was, it is a fraud that belongs to every passport. His travel documentation described him accurately and included effective biometric indicators of who he is, including a photograph. The passport is true and authentic in its representation of Udezue in every respect. It just happens that this particular instance of documented identity comes from outside any recognizable source of authority. As such, it demonstrates the sometimes-tenuous chain of identification supposed by biometric security. At the border, at least, biometrics are only helpful in verifying identity and can do little to verify the truth of the institutional memory on file regarding that identity. As a technology of matching and verification, it can only confirm whatever truth has been recorded and deemed official.

Consider the similarity between this episode and Sid Hill's account of traveling on a Haudenosaunee passport. In the eyes of the border agents who refuse to recognize these documents as valid, there is little difference between them in the sense that each one similarly breaks with a given understanding of institutional authority and who has the right to produce documentation that affirms one's identity. In each instance, the right to affirm identity is decided not by oneself but by an agent of the state. The obvious and immediate differences are that one is an attempt to travel without legitimate international documentation and the other is an affirmation of sovereignty upon the part of a recognized indigenous nation. Udezue's passport conjures up an entirely invented and nonexistent nation, while Hill's passport is properly authorized by the Iroquois Confederacy. Yet each reminds us that verifying one's identity is not something one does freely and individually, just as the state that issues the passport cannot control how it will be recognized. Identity verification initiates a set of relations that rely not just on authority but also on the recognition of authority. Therefore, an illegitimate source can be deemed authoritative, just as a legitimate source can be deemed inadequate. The scenes of recognition that make identity possible are not simply a matter of authority but rather belong to unstable relations of power that can unfold in inconsistent ways.

Udezue's case belongs to a world prior to the increase in security screening that would define international travel after 2001. In an age of biometric screening, surely such inventions would ultimately prove impossible.

RAFAL BUJNOWSKI

Or maybe not. In 2004, Polish artist Rafal Bujnowski created linked works of art entitled *Flying Lessons* and *Visa*. The performance art project began with the artist painting a portrait of himself that he then photographed and submitted as the black-and-white photograph required for his US visa application. The US embassy granted his visa featuring his painted likeness. During his visit to New York City, he took flying lessons over Manhattan. Not only does this performance demonstrate some of the limitations at the heart of biometrics, it also captures many of the fears and anxieties that motivate public support for it as a supposedly effective method of security.

Oddly, perhaps, an account of *Visa* by the Raster Gallery, which also hosts the video, *Flying Lessons*, makes no mention of the events to which Bujnowski's artwork refers. Nowhere does it mention 9/11 and the hijackers who entered the United States on legal visas or the flight training they undertook. The project is premised on an idea of mediation, or the process by which one thing stands in for another. The visa references Rafal Bujnowski, but the image it provides is not him: it is a painting of him. Then again, is a painting any less a representation of him than a photograph? The difference would appear to arise in regard to the biometric data that is available in a photograph but not a painting. But a painting may offer up a different sort of biometric record: the brushstrokes of technique and style that may mark him as a singular artist. What likenesses are acceptable and which ones are not? When assessing identity, what features are selected as relevant and which ones are ignored? Identification does not record the truth as much as it follows the rules according to which a representation can be said to be true. What matters is whether or not a representation is recognized as legitimate by the Department of Homeland Security, not whether or not the representation is actually true.

Biometric practices trade in ideas of truth and promise a firmer connection between the individual and a given means of representing and recording identity, but this remains a creative endeavor: biometrics invents something that it recognizes to be identity. Identity is always a metaphor, but a curious sort of metaphor that refers to something that possibly does not otherwise exist and which might never be adequately brought to sight. It is odd to think of official measures of identification as artistic endeavors, but they are nonetheless powerfully creative in their capacity to invent, regulate, and make legible or illegible the essence of identity.

SUAAD HAGI MOHAMUD

Suaad Hagi Mohamud was barred from boarding her flight to Toronto from Nairobi in 2009. The Canadian citizen had visited Kenya for three weeks. Upon departure, her passport was examined and an official refused her entry on the basis that "her lips looked different than those observed in her four-year-old passport photo" (Browne 78). Her passport's legitimacy was never in question, and as a result the fault lay with her. She was seen as an unidentified individual attempting to travel illegally on Mohamud's documents. This situation quickly and remarkably escalated. Apparently unidentifiable, she was nonetheless identified as an impostor and "was charged with using a false passport, impersonating a Canadian, and with being in Kenya illegally" (Browne 78). Because all of her identification confirmed the identity represented in the passport, it was seen as further evidence of impersonation rather than proof positive of her actual legal identity and citizenship.

For months, she was unable to overcome the perceived differences between a representation of Mohamud and her physical person. She submitted fingerprints in the hope of being able to prove her identity, only to find that no Canadian agency had a record of her fingerprints. After eventually receiving the DNA test she requested, Mohamud proved her citizenship via a comparison to her Canadian-born son. Three months later she was finally on her way home.

Scholar Simone Browne contends that this is an example of the ways in which, counterintuitively, biometrics can help to challenge racism because, in this instance, being able to prove one's identity via DNA challenges the power of customs agents who "seemingly determine who is in or out of place at the border in racially specific ways" (78). I cannot help but wonder about the tenuousness of this success, however. What if Mohamud's son had been born in Somalia like her and thus possessed a similarly unassured ability to verify Canadian citizenship? Or, if her son had been traveling with her, what good would a DNA test do when her son could be used as further evidence of the lengths she went to in order to impersonate a Canadian?

Mohamud herself notes the "Canadian High Commission wouldn't be treating me the way they treat me" were it not for race and religion. "If I'm a white person, I wouldn't be there one day" (qtd. in Browne 78). Is this an instance of a biased application of biometric regimes of identity verification on the part of Kenyan and Canadian officials? Is this a demonstration of how identity is lived for some but not all, never subject to trust and taken on faith when faced with circumstances in which certainty may not be possible? Is this a triumph of biometrics and the truth that can be spoken incontrovertibly by the body, or does this indicate that even biometric truth depends on being recognized as true? What may be most salient for thinking about biometrics is that this episode reveals the uneven ways in which biometrics can produce lies and truth and the ways in which each relies on so many circumstances beyond the individual physical body to generate this data. For biometric technology, the body speaks, but it does so within frameworks that govern what it can be heard to say.

CHRIS ALEXANDER

During the 2015 Canadian federal election, the ruling Conservative Party announced the introduction of a hotline staffed by members of the national police force, the Royal Canadian Mounted Police, "that citizens and victims can call with information about incidents of barbaric cultural practices in Canada or to notify authorities that a child or woman is at risk of being victimized" ("Hon. Chris Alexander"). Immigration minister Chris Alexander explained this measure by noting "we need to stand up for our values." Recalling how the wars in Afghanistan and Iraq were authorized and described to Canadians, he framed the hotline as a means of defending the vulnerable: "We need to do that to protect women and girls from forced marriage and other barbaric practices." The public immediately understood this to be a way to complain about neighbors who wore hijabs or burkas and to criminalize cultural differences.

This attempt to inflame racism selectively notices differences in public appearance and cultural practices and cynically interprets them as indicators of barbarism among some Canadians. It relies upon and reflects some commonly held assumptions about transparency and the availability of women to be seen and looked at that inform so much of public life in Canada and elsewhere. Like Craven's excursions in Turkey two hundred years ago, this measure reveals the presence of powerfully gendered biometric principles guiding social concepts of visibility and availability to inspection.

There is always a pressing need for a society to protect its most vulnerable, but I am not convinced that is what the practice of identifying and reporting un-Canadian individuals is meant to do. Instead, this biometric impulse to scrutinize and uncover others is an activity that divides a society in order to continue to not examine the ways in which some are being identified and treated as lesser citizens. It is a mode of identification that can distract from rather than train one's eyes on the very real forms of inequality and violence felt by Canadians. This was the same government that for years refused to call an inquiry into hundreds of missing and murdered indigenous women, so we should not believe its sudden concern for women and girls is anything more than opportunism. This episode reveals one of the primary operations of biometric thought: by seeing one thing and rendering it spectacularly visible, it fails to see many other details that ought to be relevant. This can mean reducing an individual's identity to what is most visible, as in instances of racism, or it may involve a stubborn empiricism, unable to see the extent to which social relations structure individual existence because such matters are scarcely visible by a fingerprint scanner.

What do we not see when we are invited to unveil so many individuals and identify them as barbaric or as victims of barbarism? To what extent do the operations of biometric thought always involve constraining the range of possible identifications? One is invited, in this instance, to see Muslim women and girls only as victims and to silently affirm, at the same time, that the normative visibility of women is all but enshrined as a pillar of civil society in Canada. Biometric thought works to make some concepts obvious, such as the moral goodness of visibility, and others much less so, such as the ways in which those same operations of visibility may subject women and girls, almost exclusively, to aggressive social control and objectification and the ways such practices have become normal within Western patriarchal societies. They are so normal and so unexceptional, in fact, that they can be deployed in order to identify another practice of regulating appearance as the only countable evidence of discrimination against women.

Biometric thought can involve assumptions regarding openness that structure public life. This means that the force of biometrics comes not just from the powers of arrest, detention, or deportation that characterize life at the border; biometrics may not always involve the minimal technology of a piece of ID but may instead work upon individuals and a society at the level of social norms that police how one appears and what one understands to be proper and improper forms of public appearance.

MARY SHELLEY

Mary Shelley's *Frankenstein* (1818) is perhaps the novel *par excellence* for biometric thought. One may not expect to find expressions of a technology as advanced as biometrics evident in a novel written two hundred years ago, but her novel is powerfully concerned with topics such as physical appearance and how social perceptions create reality, as well as what it means to identify another and how those identifications guide understandings of ethical responsibility. The story of an abandoned creature, perceived to be a monster by all who see him despite his gentle nature, who travels the globe in search of somewhere to call home sounds like a novel with ready purchase on the present and much to add to present discussions of migration and abandonment.

The novel has long been read as a cautionary tale that shows how technology can appear to operate according to an inevitable logic, as if beyond anyone's responsibility. At the same time, this is also a novel about the consequences upon individuals of technological endeavors. When we speak of biometrics, we have a topic that seems blameless for any of the harm it does or the consequences that follow from its particular ways of understanding the world. The harm is always elsewhere, not quite the fault of the technology. My goal is not to assign blame as if to say, yes, this technology is really doing this and we have caught it red-handed. Rather, I think it is important to understand how we have come to view technology as neutral and what we lose when we readily accept the narrow perspective necessary for this claim. It also means developing and refining ways of thinking that can identify what biometric technology does and what counts as "doing" in this context.

There may be no stronger account than *Frankenstein* of what it means to see and not see another and of all of the ways in which cultural norms intervene to regulate how individuals appear. One of the most remarkable scenes occurs when the creature views himself for the first time and discovers what he looks like reflected on the surface of water. This is a deeply affecting scene, and readers pity the creature because he is literally unable to see himself except as a disfigured monster. With this scene, Shelley shows the reader the influence of social norms upon how one processes what one sees, such that the creature can only see himself as if through the eyes of others (and come to think of it, that may be literally true for this strangely assembled creature!).

The creature is mortified to discover he is a monster and is unable to see himself as anything but his exterior form. Why is the face of the creature a vision of terror? Why can others not read his childlike vulnerability in the lineaments of his disfigured and admittedly far-from-normal face? What explains this emphasis on physical appearance and the inferences that others draw from it, such that they see a monster rather than a being who suffers? Shelley writes at a moment in time when European culture is developing a powerful new understanding of the ways in which the exterior of a person reflects inner moral and psychological qualities. This was the beginning of a biometric culture trained to see the body as an expression of the essence of a person that continues today in advanced biometrics.

Obviously fascinated with countenances and the product of a culture obsessed with reading faces, *Frankenstein* offers a sustained response to an age given over to the fantasy of being able to read people as if they were books. The novel challenges many of the assumptions of a biometric culture of face reading popular in Europe at the turn of the nineteenth century and which has continued to this day. It charts what it means to live in the wake of such assumptions. In a sense, biometric thought aspires to make everyone into the creature: separated from the social circumstances and contexts that illuminate how they have come to be, defined by visible surfaces, and subject to outside recognitions of one's particular essence, which is to say how one might be assessed according to existing norms of intelligibility. Such a perspective on the novel is perhaps particularly important for readers now, given that many continue to live in societies in which duty toward one's fellow creatures is routinely policed on the basis of the apparent familiarity or strangeness of another's appearance. What kind of viable social existence can be secured for all when so many beings are identified as lives that do not matter precisely because they fail to be sufficiently normal, acceptable, or *like me* in shape or appearance?

One of the great accomplishments of Shelley's novel is the realization that monstrosity is what is consistently identified in another but turns out not to exist. The creature is far more human and humane than he appears, and even if Victor's failure to care for the creature is itself monstrous, he is nonetheless also a shattered individual who has trouble caring for anyone. If biometrics tends to conjure up migrants as suspicious threats to the nation, then perhaps the lesson to be drawn here is that a stranger is always far more human than the biometric regimes for establishing identity may be ready to perceive.

The narratives I have recounted here each represent how biometric thought mediates individuals' experiences and knowledge and shape reality every minute of the day. Despite its attention to the particular differences that make an individual unique, biometric identity verification sometimes seems abstract, as if it does not impact real human lives in substantial ways. As all of these individuals know, biometric technology and biometric thought establish part of the ground upon which they live. These are never the only factors that shape their lives, and perhaps the influence of a biometric logic is sometimes lost because the narrow practices of biometric authentication seem benign, as if belonging only to an occasional world of procedural routines.

Lauren Berlant writes of the figure of "slow death" in order to capture the ways in which some social crises do not get recognized as such or receive little attention compared to other, apparently more obvious, crises. How we speak about the realities we encounter matters and, strange as it sounds, speaking about something as a crisis may add to the problems we hope to address: "Often when scholars and activists apprehend the phenomenon of slow death in long-term conditions of privation, they choose to misrepresent the duration and scale of the situations by calling a *crisis* that which is a fact of life and has been a defining fact of life for a given population that lives in crisis in ordinary time" (101). The recent civil war in Syria that has displaced approximately five million citizens to Lebanon, Jordan, Turkey, Iraq, and Europe may be exceptional, but similarly precarious efforts to migrate have become all too regular for many others. If now is the time to think clearly about such matters, it is not because the world is facing a crisis in human movement; it is because human movement has become more precarious, more tightly regulated, more necessary, and more routine.

What would it mean to speak less about a humanitarian crisis and more about how we comprehend human movement and the biometric procedures that presently police how we conceive of this reality? Biometric thought has perhaps been at the root of a great deal of slow death in which human mobility and immobility, suffering, and the desire to make a livable life may never attain the visibility of a crisis or become so unusually pronounced as to be deemed an emergency. I hope that this effort to think about how identity and movement are regulated and about the intended and unintended effects of that regulation can establish how biometric thought has, for a long time, been an ongoing way of grounding identity and framing the world. As such, biometric thought needs to be made more visible because its operations and assumptions have affected, and will continue to affect, a great many lives around the world.

One of the most powerful effects of biometric thought that this book considers is well evidenced by these individual narratives. Biometric thought promises to do what I have done in this chapter: it names distinct individuals and tells a unique story that identifies one person. It promises to identify people in such a way that they exist as discrete individuals, attaching identity to measurable information collected from the physical existence of the person.

Yet if we take a second look at the narratives I have included here, we see that in each case, identity is social. Each individual relies on others in order to be who they are, and they cannot, in themselves, provide the basis for their identity. For example, Mohamud could not prove her identity without reference to her son, and there was no need to do so except that government officials had the power to disqualify her own assertions. The Canadian electorate learned that national identity is defined not by its essence but by its difference from so-called barbaric others. Bujnowski's examination of biometric thought was a performance that depended upon being watched by gallery patrons and by referencing prior terrorist attacks. Udezue's deception was only remarkable because he was welcomed into so many nations on a baldly fraudulent passport. Even Kiyani's trigger lock was an invention rooted in an acknowledgment of the presence and importance of those who might wield a gun owned by another. Suriya's experiences as a migrant laborer in Qatar was defined by his isolation from friends and family and by conditions that forced such an existence upon him. Laymon's account of his ID and the ways in which it protects him from harm was especially startling when placed next to the regular vulnerability to violence experienced by African Americans around him, individuals whose most marked difference is only that they cannot similarly produce such a powerful piece of plastic. Sid Hill argued compellingly that his identity depended on his Haudenosaunee culture and national status; it was not something that belonged to him and him alone. Elizabeth Craven's experience of the Eastern veil depended upon her own social experiences as a woman gossiped about. Frankenstein's creature came to know himself as an effect of the perceptions of others. Harun al-Rashid's practice of circulating in disguise amongst his people made it clear that the social recognition of oneself by others has the power to make and unmake individual identity.

Biometric identification promises to attach identity to the physical body and make it verifiable in order to test authenticity. But it does much more than this. It establishes a fantasy that identity is attached to a singular individual, which could not be farther from the truth, as each of these different individuals knows. Biometric logic invalidates the ways in which culture and customs structure existence. What are the effects of making the social existence of identity disappear from view and replacing it with a strident independence from others? What does it mean to create, as an expression of advanced technology, conditions that seek to annul what is most human about existence? We need others and depend on others for all that we do and to understand who we are. Languages are social creations; our societies can, even if they sometimes fail to do so, support one another and make life more livable; every sense of ourselves depends upon social structures and long-established concepts and culturally specific ways of thinking. An idea of the self as something entirely contained within a body not only ignores this reality but also conjures a mode of existence in which we are encouraged to believe that we can do without others and without the social supports that make life possible.

In many of these examples, biometrics seems to be a state power and references an ambition to regulate how people and populations move. These initial examples already suggest, it is worth noting, that this power is not absolute, despite how consequential it is. Indeed, part of what I want to document with these examples is a sense that biometrics is more than the embrace of a sovereign or national authority over one's body and identity, as if the demand to identify oneself is simply a matter of submitting to the all-encompassing power of the state to decide one's fate. What we see here, instead, is that individuals navigate and, in a sense, negotiate with the demands and normative expectations of biometric thought. While biometrics may sometimes seem to be the means by which to secure the borders of a state, it may also open up new forms of evasion and new ways of understanding what evading the impulses of biometrics might mean, given all of the varied effects associated with this technology of identification.

We ought to abandon, from the start, any idea that biometrics is a technology of absolute control or exclusion. It may often wish that it was, but there are significant gaps between what it desires and what it accomplishes. We need to also abandon the ideas that biometrics is simply unwelcome and always repressive in nature and that it is uniformly effective for and experienced by all individuals.

Insisting instead that biometrics entails complex and contingent relations of power makes it possible to account for the very different ways that biometrics operates in different locations, for different purposes, and among individuals whose experience of state-sanctioned measures of securing and confirming identity may be quite distinct. Biometric thought is always characterized by its uneven and inconsistent application across the globe and with varying degrees of consequence for those whose lives are somehow conditioned by its operations. Even if all the individuals whose stories make up this chapter share interactions with biometrics, their experiences and the meaning of those experiences differ widely. Biometrics does so much more than simply identify individuals; it possesses a varied and social existence, and its effects may depend significantly on the circumstances within which it operates.

I have gathered together a number of quite different encounters with biometrics in this chapter. Some are drawn from everyday life and document how such interactions with technologies of identification affect people directly. Others are fictional, sometimes remote from the present and sometimes directly relevant to it. My own sense is that fiction and reality are not necessarily an especially powerful division for biometrics, given the way it produces and relies on particular fantasies of identity and how they can be located in the body. Do these accounts offer a human-centered narrative, I wonder, that is an alternative to a techno- logical mode of apprehending the human? These are long-form expositions, not data sets, after all. I am fascinated by such an opposition, but I do not see this as the basis for my argument. These human encounters with biometrics do not come to matter, for me, because they speak a *living* truth as opposed to the somehow less true *representation* of oneself that emerges from biometric technology. I might be tempted to assert an opposition between impersonal technology and these human- focused encounters because it is in the realm of the human that we can most read- ily and immediately see that identity involves forms of narration. But this is not something exclusive to human experience, either. We are always narrating identity, and this is true for biometric apprehensions, too. Biometric technology is itself a mode of narrating identity, not something opposed to it. It is a very particular form of storytelling, to be sure, with a unique beginning and end. Its story can vary internally, depending on the matching requests and the data sets, just as it can focus on different features of a person, be told for different reasons, or to dif- ferent audiences, with different results and implications. Human experiences are always going to be mediated, and while binary code is a different sort of media- tion than storytelling, I am not convinced that one is true and the other artificial. That seems to misunderstand the ways in which the raw material of reality will always be subject to mediation of some sort, even if it remains pressingly impor- tant to notice the differences between one form of mediation and another. It may also be worth adding that some of the accounts in this chapter take on a sense of reality and gravity precisely because they are stories or scenarios artfully told. Rhetoric can make things compelling, just as reality can seem desperately inau- thentic if poorly narrated. When one thinks of biometrics as an art of telling a story about identity, presence, and existence—complete with its own rhetoric and conventions—one can begin to see what it does more clearly.

What I want to stress, more than anything else, with these examples is that bio- metric thought implies ideas of human identity, and we need to be alert to how it regulates and produces these ideas. Writing about the tension between oneself and the alienating experience of submitting biometric data in order to verify iden- tity, Irma Van der Ploeg arrives at an impasse: "So we are stuck with a riddle: how can a biometric identifier be both identifying and not say anything particular about you?" (76). Biometric authentication does not feel personal, in other words, even if it apprehends someone at a level that is more attuned to the individual than ever before. While an individual's identity might rest on personality, or on a given capacity to care about others, or on a set of accomplishments, or on how one fits within certain social categories, it tends not to rest on a pattern of blood vessels within the eye.

Biometric identification reinforces the idea that I coincide with my body. And in this sense, to be unsettled as Van der Ploeg is by the tension between what is recognizably and unrecognizably oneself is already to have forgotten ways of thinking about identity that are not limited to one's own body and its operations. One's sense of self might depend, for example, upon the force of social attachments to a network of friends and family first and foremost. Or it might depend upon a powerful connection to a given part of the world and to the rhythms of that land. These are modes of understanding oneself that depend upon what is beyond oneself as the basis for identity, on experiences of proximity and affective relations, not sheer individuality and bodily separation from everything else. What may be the strangest thing of all about biometrics is how readily it helps us forget that identification entails what is other than oneself, what is strange, and all that is outside of oneself.

In the next chapter, I consider what it means to speak about biometrics as a mode of thought and how this curiously social side of biometric technology operates, including how it frames reality and polices how we think about human mobility.

2 · THE SOCIAL LIFE OF BIOMETRICS

In the novel *Preparations for the Next Life*, Atticus Lish tells the story of two migrants who make a life together and try to understand what it means to live as foreigners in New York City. Zou Lei arrived as an undocumented migrant from China, while Brad Skinner is an American citizen who has returned from war in Iraq. Neither one finds it easy to be at home in oneself, and neither knows exactly how to live in this foreign city. The novel reminds readers that identity depends so much on those around us, who make it possible to grow and change and come into a sense of ourselves, as well as on social circumstances that can variously sustain and undo us.

Zou and Skinner lead lives defined by conditions they do not choose and cannot control, conditions that threaten to dispossess each of them differently. Skinner is debilitated by his PTSD (post-traumatic stress disorder) and the inability to have done with modes of living necessary for survival in war that, in turn, make life almost unendurable outside of combat. Zou finds herself abused by employers, marginalized by fellow migrant co-workers who are unable to understand her Uighur language, and preoccupied by fears of deportation or another period of immigration detention. Both have been transformed by their travels around the globe, and both continue to struggle to understand who they are, especially because they find that the recognitions of others entirely fail to comprehend their experiences. Taken together, their experiences reveal unexpected moments of convergence and provide a portrait of what it means to live in the wake of various social and official practices of human identification that intentionally and unintentionally attenuate the complexity and richness of their actual existence. For Zou, this is sometimes manifest in the form of a constant fear of being discovered:

> She dreamed that ICE agents came to Chinatown in the new white Homeland Security trucks and piled into the mall and closed off all the exits, shut everything down, and started checking all the workers with a biometric scanner. The agents

put them all in a line, made them raise their arms, and scanned their hands and retinas. Polo and Sassoon were cleared and allowed to go, but the scan caught everyone who wasn't legal; the Mexicans and the illegal women. The agents made them lie down on their stomachs. There was nowhere to run and they were going to get her. (179)

To appreciate that identification exists as a process that can dispossess individuals, not just confirm identities, the novel recognizes the human narratives of desire and anguish that come with living a life that fails to be verifiable. For Zou, this means living in fear of immigration officers and constant anxiety about what others might detect in her appearance. For Skinner, this means that what is most definitive about himself is what he least wants to acknowledge or claim: his own suffering and mental illness. Both Skinner and Zou are jeopardized by the ways in which they are recognized. Zou lives and works under threat of detection and thus without any rights or protections. Recognized to be back at home and discharged, Skinner is abandoned by the military that has transformed the way he thinks and experiences reality, and this transformation has made everyday life unfamiliar and so much more than he can handle. Significantly, the novel confronts the reader with the idea that their different legal identities may have little bearing on how similarly vulnerable they are or just how much that vulnerability constrains their lives.

Preparations for the Next Life shows that biometric identification needs to be understood to affect individuals differently, such that Zou's dispossession as a Uighur Muslim forced to flee China is legally recognized in only a limited form that sees her presence as illegal. Meanwhile, Skinner cannot have his dispossession recognized at all because he is seen to be at home and not the alien resident that he feels himself to be. As the novel powerfully contends, there is no proper experience of human movement and its effects upon an individual. Regimes of official identification affect each of them differently, and these regimes equally transform their lives such that one cannot claim, in either instance, that identification is *just* identification and lived identity is an entirely separate matter. To be paperless means that Zou knows how little she can do to protect herself from exploitation. To be identified as a civilian, not a soldier, becomes impossibly fantastic for Skinner because this alteration in his relationship to the state never adequately acknowledges just how estranged he has become from civilian life in the wake of war. Indeed, as Lish shows, there are so many effects that radiate from official practices of identification—including whether they are made or withheld—and these may leave one more vulnerable and subject to fewer protections and supports.

Such an account of the challenges that come with how one might be recognized—whether as an undocumented migrant or a war hero and veteran—demonstrates that biometrics might refer to much more than a technological and official operation despite the ways it promises to link identity to documents and ID cards to a living body. One might immediately insist that biometric practices are rooted in discrete and neutral activities of matching a person to the official record on file, but what this novel insists upon is that individuals live precisely in the gap that opens up between the person and the record and that the presumptions of matching one to the other can transform existence in consequential ways and even make that existence illegible. To be subject to such practices is a physical experience that can dispossess an individual and make one's embodied existence feel strange and ill fitting precisely because it is recognized in a manner that is unfamiliar and even impossible to comprehend. Biometric practices are rooted in physical existence, yet they are nonetheless unwilling to acknowledge the full range of how individuals live and breathe. Verifying identity also means deciding what parts of identity matter, and that experience is shown to be profoundly alienating in this novel. More, *Preparations for the Next Life* offers powerful reasons to be skeptical of the idea that biometric identification is more real because it is attached to the physical body and its markers. Identification always involves more than just the body. It entails an entire world of living relations and experiences, forms of recognition of who one is that are not adequately accounted for by biometric practices and which are never entirely separate from their operations.

If a human experience of biometric identification entails all the pressing anxieties and consequences considered by Zou's dream, there is another, different, social existence of what I call biometric thought propagated by biometric practices and widespread perceptions regarding its operations of identification, and this is what I would like to examine in this chapter. The social life of biometrics is born of the assumptions and organizing perceptions that define biometric procedures, but it involves more than this, too. Biometric thought includes but is not reducible to the human experience of living under regimes of biometric identification. It is a way of thinking. It is a mode of comprehending, organizing, and regulating how identification and mobility intersect in the form of access control. It is likewise a mode of sorting and organizing reality as if according to visible and obvious features: for example, one is undocumented, another is not; one looks nationless, another does not. It includes forms of knowledge that structure one's understanding of human existence in the world, knowledge that is almost always imbricated in forms of power and the regulatory capacity to insist on specific modes of making sense of the world that can silently exclude important parts of how one exists.

Biometric thought is not *just* a mode of thinking, however. It is made possible by the authority to insist that individuals submit to biometric recognition at the border and by the regimes of knowledge that organize and regulate the world rather than simply represent features of it. It is a mode of knowing a world of social and political relations that creates, regulates, and constrains what we can perceive about them. This means that knowledge is not simply something one has or does not have but is also something that does things in the world. While biometrics appears to simply read an already present body for whether it can match a pre-existing set of coordinates, the practice and its effects are part of a much larger social life in which biometric thought makes the world and our lives within it intelligible in specific terms and seeks to rule out modes of existence that it cannot or is unwilling to acknowledge. As such, biometric thought conditions how we think and what we are encouraged think about when we are invited to contemplate modes of identification and the purposes for which they are deployed.

Before proceeding, I want to briefly clarify one detail. I describe biometric thought as an effort to regulate access rather than as something that restricts freedom because I do not find such an opposition between freedom and control to be helpful. Human movement, for example, is never simply free. It is always conditioned in some way, made possible or facilitated in some way. The movement of the refugee or the migrant is not free in an obvious sense: it is enjoined by particular circumstances including dangers and neglect that imperil one's ability to survive. The choice to leave one's home under such conditions and the choice to confront the unknown challenges that follow are hardly choices at all. Human movement is almost never radically free, motivated and structured as it is by any number of ideas of *home* and *away* and by financial considerations that constrain movement as well as make it vitally necessary. Biometric practices do not curtail a migratory impulse that would otherwise be unchecked. Such a thought is almost always offered as part of a rhetorical strategy of securing civilization from animal-like hordes at the gate. Movement is always constrained and defined by particular and unique circumstances. More, the supposed universal and neutral application of biometric technology tends to obscure the differences that make movement more free for some than for others.

Freedom can be worth insisting upon. It is an explicit feature of the United Nations' "Universal Declaration of Human Rights," which carefully states "(1) Everyone has the right to freedom of movement and residence within the borders of each state. (2) Everyone has the right to leave any country, including his own, and to return to his country." Yet, freedom has other meanings, too. As scholar Mimi Nguyen notes of some of its powerful deployments, freedom is often understood as a feature that belongs to liberal, capitalist states and thus exists as that which will be given, too often by force, to those who lack it (23). Under such a way of thinking and acting, war can be misremembered as the gift of freedom, and refugees become individuals who should be "grateful for freedom" they have acquired in a new life (180). Biometrics is not well designed to comprehend the complexity of such deployments of freedom, and it may not be able to remember all of the conditions that make movement possible and impossible, necessary or desired, feared or derided. Such conditions are not well named by freedom or its absence. As a result, I have little to say about freedom of movement. Instead, I hope to consider how ideas of movement themselves arrive, what they intersect with, and in what ways various social forces seek to regulate both their operations and meanings.

As a promise to identify individuals according to visible signs associated with the physical body, biometric thought began well before recent advances in imaging technology. It has a history and has developed into a social force over the past two hundred years. Whereas biometric practice can respond "true" or "false" to a request to match a person to a record of identity, biometric thought has helped to define the rules according to which identification is possible, as something that can be true or false and under what terms it will be recognized as one or the other. The world is not limited to the things we name and identify; instead, naming and identifying makes possible a world of things. Where and how we separate things and understand one thing to be discrete and unrelated to another matters. To the extent that biometric thought does not just record things that exist in reality, what constellations of force and influence help to define biometric thought's tendencies to see reality as it does?

The identification of a human being as "undocumented," for example, is not a matter of describing someone who actually exists. It is a mode of insisting upon the idea of identity documents as the ground for existence. When one speaks about "illegals," one organizes the reality of individuals by prioritizing one particular feature of their existence: their immigration status. This is a mode of representation that dehumanizes individuals by reducing them to a social category that combines moral and legal force: they who lack papers are not supposed to be here. Worse, the term *illegal* has often been the first step in authorizing violence and neglect by refusing to recognize some individuals as full human beings. This designation efficiently places some beyond dignity and responsibility. Further, "they" are seen as a collective. Incredibly different and varied lives are effectively reduced to a stark commonality despite the innumerable differences that brought them to cross a border, led them to be undocumented if they are, and the different nightmares and dreams that can inspire human movement.

By identifying another as illegal—*that person, over there, is an illegal*—one might also implicitly mark oneself as legal or seek to be recognized as part of the national fabric *as opposed to them*. What is the basis for this claim? What does this claim remember and forget by insisting on such a calculus of punctual identity? By virtue of the accident of my birth, I can claim a legal status to exist here, in this country. Yet my existence in the unceded traditional territory of the Okanagan Nation depends upon a history of colonialism that includes the violent dispossession of indigenous peoples from their culture and history, the theft of territory, a history of residential schools that sought to decimate indigenous cultures by insisting on the rightness of European civility as well as the racism and prejudice that still exclude First Nations in subtle and overt ways from social life within the Okanagan Valley. My status here is the product of settler colonialism, but my society has decided that some of us who live in the wake of such migrations are legal citizens, despite never receiving documents of permission from those who were already here. Likewise, this decision now criminalizes so many of those who arrive with just as little as my ancestors did.

I highlight the ways in which individuals are officially recognized to belong or not belong, to be human or less than human, because this is one basis for how biometric thought polices and produces a sense of reality. It is a mode of recognizing the world that builds upon fictions of official identity that highlight some details as relevant and selectively ignores others. It is something we can examine and assess in order to develop new ways of organizing reality that do not authorize and encourage harm but instead ask questions such as why someone might risk so much to come here, as those before me did, and what conditions make this necessary? Or what is lost by a recognition that one exists first and foremost as a mobile unit that can be separated from family and one's culture and land, a species of biometric thought that was weaponized by residential schools that sought to sever the rich social ties to family and culture and place that define many First Nations?

One way of thinking about biometric thought and its social life is to consider what it is asked to accomplish. It is said to provide border security, guarantee identity, protect individuals from terrorism, prevent identity theft, secure any number of commercial transactions, prevent or dissuade fraud, transform unknown threats into identified and known individuals, institute principles of transparency and visibility, see through disguises, penetrate anonymity by recording and verifying a digital or physical presence, represent a physical body such that it exists not as an aging and deteriorating thing but instead with the stability of a permanent record, and operate discretely at a large scale. Biometrics cannot possibly do all of this, especially when we remember that it is a technical means of enrolling an individual via a representation of a physical feature such as a fingerprint that can then be verified later in order to confirm identity. The gap is considerable between what we want that technical procedure to accomplish and its actual operations.

Biometrics is a mode of organizing reality that is always within the realm of fantasy and ambition, producing a persistent promise of transparency and openness that may not be possible or reasonable. Fingerprinting becomes the answer, under biometric thought, to identifying criminals, then, rather than solving the much more difficult matter of why the United States has 5 percent of the world's population and 25 percent of its prisoners or why African Americans are routinely asked to identify themselves and are incarcerated at a rate of six times higher than white Americans. Iris scans promise financial security, rather than thinking about how the economy functions to dispossess so many, globally and domestically, who are desperate to survive. Smart passports ensure regulated human movement, rather than reflection on the nature of foreign policies that seeks to solve complex geopolitical tensions by dropping bombs or overthrowing governments.

Facial recognition secures identity, rather than seriously addressing what it means to live together on a finite and warming planet that is primed to produce ever more climate refugees. Biometric thought dreams of simple technological solutions that, if they see larger circumstances at all, reduce those to matters of matching and verification. And to say that it is dreamlike in the way it functions does not in the least diminish its effects upon individuals and communities.

Some will think of biometrics as a dream of the future, a technology that is always on the cusp of perfection. Many associate biometrics with advanced technology such as palm reading touchscreens or the iris scanners one sees in film and television. Such technologies exist and may become more common in the future, but this is not the dream I have in mind. As a dream of the present, rather than the future, biometric thought captures a sense of social desires and ideals, technological optimism, and produces ways of understanding reality right now. Biometric thought may even render it more difficult to see and envision alternatives to biometric impulses of identification and why alternatives might be desperately needed at the present. Less an indication of the arrival of the future, biometrics may herald the impossibility of escaping presently sedimented ways of thinking.

The effects of biometric thought may be even greater than the ambitious purpose such procedures are asked to serve. For example, as part of a coordinated security strategy, biometric thought identifies strangers as individuals who might do harm to others. This also means that it helps to produce an idea of oneself harassed by the existence and arrival of strangers and normalizes the idea that one ought to be free from social relations and unchosen bonds with others. It is a way of thinking about the world and those around us, whether close to us or on the other side of the globe, that seeks to recognize individuals as potential threats before it sees them as fellow beings who have comparable fundamental needs or as individuals to whom I have obligations and responsibilities. Not only are "we" made vulnerable to a specific threat from outsiders, this logic also leads one to infer that the stranger is the most dangerous of all figures. Consider CNN's data on the United States from 2001–2013 that showed for every one death by a terrorist, one hundred died by firearms (homicide, suicide, accident), with a total of 3,380 and 406,496 deaths, respectively (Jones and Bower). Yet within the national imagination, terrorism from outside the United States is the greatest threat to safety and security (to say nothing about the considerable history of domestic white-supremacist terrorism). How many deaths will it take before some recoil in fear not from strangers but from family members and friends who are overwhelmingly the perpetrators and victims of gun violence? How reality is framed and the lived experience of reality may not align at all.

My insistence that biometric thought frames reality and produces a range of effects offers a very different scope for analysis than existing considerations of biometrics as a means of establishing state surveillance. While biometric procedures might exist within a larger web of governmental monitoring, biometric thought refers to much more than surveillance. It is absolutely part of a surveillance apparatus, but it is a mistake to understand that as an explanation of its only or even its most profound purpose. If we understand the primary function of biometrics to be one that confirms identity in order to regulate access, then discussions of individuals watching and being watched address unevenly, if at all, many of the most important features of biometric thought and its preoccupations with migration, borders, identification, legibility, transparency, and security.

Biometric thought may lead to forms of surveillance, but given its powerful reliance on normative conceptions of identification and how its ideals are relayed and intensified by diffuse points across social and cultural life, it cannot be assessed solely as an empirical phenomenon concerned with "a broader system of connected activities" such as data collection and monitoring (Marx xxvi). Much of the social life of biometrics involves ideas and concepts that are established by practice but cannot be said to exist in practice alone, either.

The writings of Michel Foucault are often cited as a starting point for surveillance studies, yet I see clear differences between this newer archive and his ideas. Foucault's analysis in *Discipline and Punish* moves from technologies of surveillance employed in prisons to consider a wider social world of disciplinary correction in which individuals behave as if they are being watched. It is a model designed to explain how it came to pass that "the useful administration of men first appeared" as a central problem in "productive" democratic capitalist societies (303). Key to Foucault's theory is the idea that there are consequences for acting in improper and deviant ways. The most aggressive expressions of police powers are rarely necessary because individuals largely self-monitor, having internalized forms of normative self-regulation and social ideals of desirable and appropriate conduct. They discipline themselves as if someone is watching, one might say. Foucault is working with a sense of power here that suggests social forces can be generative rather than punitive. The power to train, observe, assess, correct, and discipline are powers that fundamentally shape social life to this day. This is a social theory of individuation that sees the historical development of the modern individual in the operations of a "specific technology of power that I call discipline" (194). Such disciplinary societies are defined by widespread social norms that are inculcated at various points across society, including popular media, education, as well as work, and it is hard to imagine a nation on earth that could not fit such a description at present. To be an individual is to recognize that one lives in a social world of normative expectations and that failure to pay heed to these norms may produce greater or lesser disciplinary consequences, ranging from the mild to deadly. Social discipline is effective because it is pervasive and constant rather than singularly punitive. Instead of a single police body regulating proper behavior by force, the operations of normative social discipline are relayed by teachers, employers, parents, media, and so on, and reinforced by individuals themselves. What Foucault understood better than most includes two key ideas in this context: (1) that order is an effect of subtle, almost invisible, and infinitely diffuse operations that rarely take the form of a Big Brother–like edifice of direct social surveillance and are instead informal and institutional, supported by norms and conventions and not just regimes of law and physical monitoring; and (2) that the largely productive and compliant individuals of disciplinary society are produced as an effect of these measures of social normalization; they are not pre-existing entities who are then subject to surveillance. Individuals are shaped by normalizing forces, and these forces surround them from birth until death. Foucault was not describing totalitarian states, in which monitoring was pervasive, but free and democratic societies that, starting at the end of the eighteenth century, created social mechanisms to ensure that freedom did not lead to idle dissipation or lawlessness and that human rights did not lead to a desire to be free from unwelcome and exploitative economic relations. Alongside the law and rights and freedom came a social world of normalizing practices that ensured citizens were productive and largely docile: "Historians of ideas usually attribute the dream of a perfect society to the philosophers and jurists of the eighteenth century; but there was also a military dream of society; its fundamental reference was not to the state of nature, but to the meticulously subordinated cogs of a machine, not to the primal social contract, but to permanent coercions, not to fundamental rights, but to indefinitely progressive forms of training, not to the general will but to automatic docility" (Discipline 169). Where Foucault attended to what it means to live as if one were being watched—a world in which expectations born of the military dream of normalization, regulation, and disciplinary correction have been achieved—surveillance studies look largely at empirical surveillance and its effects upon individuals whose freedom is curtailed in some fashion. This implies a radically different starting point, then, from Foucault's work; it implies that one is a free subject whose rights are then curtailed by a state that monitors one's email and texts. Foucault, by contrast, noted that a disciplinary society produces individuals who might never need to be monitored because they have internalized the norms and ideals of their society so thoroughly.

So, while scholars in this field might utilize a language reminiscent of Foucault, when they speak of forces of social control shaping a society and how one lives in the world, they are much more empirically minded. Bauman and Lyon assert "the next generation of drones will see all, while staying comfortably invisible—literally as well as metaphorically. There will be no shelter from being spied on—for anyone" (20). Ball, Haggerty, and Lyon note that surveillance has become both more visible and less visible in recent years. They suggest that "as we go about our daily lives it is hard to miss the proliferating cameras, demands for official documents and public discussions about internet dataveillance" (3). And yet, "there is a curious invisibility surrounding these practices," such that the nature and use of these operations of surveillance are "opaque to all but a select few insiders" (3). For these scholars, surveillance is a literal operation: there are cameras and people are watching. I am less interested in assessing empirical biometric measures of identity verification and access control, and more interested in how they invent, as an object of regulation, an identity that they can then verify. Rather than work with the idea that there are individuals who then have their identities confirmed by biometric measures, my analysis extends Foucault's thought by arguing that biometric measures help to produce the very notion of identity that they seek to regulate.

Rather than approach biometrics solely as an empirical apparatus that has been overlaid on top of society as a particular form of monitoring, I consider how a concept of biometrics relies upon and invigorates social desires and norms and thus in turn shapes the very society of individuals it claims to make legible according to biometric techniques. Biometrics offers a way of thinking as much as a set of tactics of observation made possible by technological advances. Thus, while my consideration of biometric thought is likewise interested in visibility and invisibility, it also asks: How does a concept of biometric legibility police what can and cannot be understood to form the basis of a person? If identity is grounded in the physical body, in what ways do other features of identity, such as one's social attachments and attachments to the land, cease to be legible as equally powerful and determining features of identity? Such a question contests a practical view of biometrics that follows its gaze and sees only physically present individuals. Why should sight condition what we can consider to be relevant? Similarly, an emphasis on surveillance tactics is likely not to hear such a strange query regarding the normative judgments about the nature of identity that are instituted and reiterated by biometrics. Focusing only on what is empirically available, in other words, may not get us to the questions we need to ask.

Finally, I share scholar William Walters's concern with some expressions of surveillance studies that produce a perhaps unintended awe and reverence for the very thing they study. Walters identifies this as a form of the sublime, a term Romantic-era poets and philosophers reserved for the human experience of impressive and overwhelming scenes, such as the sight of a mountain's grandeur or the vastness of the universe. Arguably, this affective response to the sublimity of state control is the least productive aspect of surveillance studies. It is animated by a sense of "technological sublime" produced by a "disposition which makes modern surveillance technology appear as awesome in its power and capability" (59). I suggest this perspective is not helpful because it can actually produce the very docility it documents by implying that technologies of social control are overwhelmingly powerful and supple and thus resistance is always already a lost cause. I am unwilling to simply accept what I find unwelcome and unjust about the world and its operations, and I am especially unwilling to accept such an impoverished notion of resistance and what it means to live against, despite, or without such measures of social control and with so little thought as to all the ways in which people are doing so. As individual accounts from the previous chapter show, the practical and philosophical operations of biometrics are defined by their contingency, variability, and the ways in which they intersect with other power relations. To suggest that biometrics is uniformly effective or that it consistently achieves the same ends is simply not reflected by the evidence, even if it is plain that it aspires to regulate access by facilitating identification and authentication.

Against this technological sublime, I contend that biometric thought illuminates a number of pressing social issues and frames them in particular ways. And if it makes matters thinkable in specific ways, I am fascinated by how we might yet conceive them otherwise. The reach of biometric thought is not total, especially if we can begin to understand that any conception of biometrics depends on how a society recognizes its operations and its effects.

Biometrics has a social life, I insist, and is not reducible to its empirical operations. Biometric thought frames reality by recognizing some matters and not others, and these recognitions are supported by numerous social norms that likewise condition how we perceive the world. By making this claim, I directly challenge those who would say that biometrics is a neutral technology—and that it is neutral, moreover, because it is a technology. Biometric thought is distinguished from either proper or prejudiced applications of an otherwise neutral technology. It captures the ways in which the practice of biometric authentication participates in and produces larger social assumptions regarding the nature and locations of identity, alongside the varied effects of regulating identity upon individuals and social structures. Biometric thought polices how we recognize individual and collective existence and what one can know about the reality within which one lives and breathes. It is not merely a set of identification practices alone, as some might suggest by casting it as a neutral technology.

The notion that biometrics is neutral and machinelike in its indifference to subjects and operators might lead to the following assessment: biometrics can claim something no border guard can be sure of because it does not judge on the basis of skin color or ethnicity and, as a result, may be preferable to potentially biased human judgments. Shoshana Magnet has considered this perspective and notes that one of the benefits touted about biometrics is the belief "that new technologies will circumvent forms of 'race thinking' and racial profiling by replacing the subjective human gaze with the objective gaze" of a biometric scanner (25). Yet biometric procedures are rarely fully automated and necessarily involve human administration and human design. This can lead to at least two overlapping conclusions. One is that biometrics is simply a tool, and like any tool it can be used well or ill depending on the person wielding it. The second is that the tool itself has been prejudiced by its design and works in a manner that reflects those assumptions. Magnet shares examples of biometric research that read more like fringe science, in which biometric categorizations of race rely on the discredited pseudoscience of anthropometry in order to assert that race is a physical feature that might be measured in the lines of the face and the angles of the head (43). Similarly, who is subjected to special biometric scrutiny may reflect social priorities rather than objective judgments. Collection of advanced biometric data by the U.S. Department of Homeland Security was required for individuals arriving from approximately forty Arab countries in the years after 9/11 (47). These selective measures remain in effect for immigrants and refugees from Iraq and Syria (Worth).

These examples are not necessarily an indictment of an ideology of technological neutrality. Instead, they are an indictment of the administrators, technicians, and engineers who have failed to be as neutral as their tech. And in cases where the science still draws upon "assumptions that categorize individuals into groups based on phenotypical markers of racialization and used in the implementation of programs specifically aimed at racial profiling" (Magnet 43), the fault is ultimately a human one for developing a tool that assumes a norm regarding the presentation of particular facial features and how that correlates to social ideas of race and assumptions about who poses a threat to national security. We simply need more enlightened and more responsible scientists, this line of thinking suggests, alongside better training for those implementing and administering the devices.

To understand how biometrics comes to appear pragmatic and impartial, it is worth considering a parallel example. Carey Wolfe has examined how instances of impartiality and neutrality function for bioethics, a field that purports to think about ethics in relation to medicine and the life sciences and which addresses controversial matters such as cloning, informed consent, genetic engineering, and the use of nonhuman animals in experiments. It is a field that quickly reduces the question of "what should we do" to one of calculation and formulae (Wolfe 95). As Carl Elliot notes, this is a field that is "constrained by the demand for immediately useful answers" (xxii); thus, what it presents are decisions to be made that can be justified as ethical. Bioethics often involves determining prescriptions and best practices that can then be followed in order to ensure ethical conduct, functioning largely as a set of rules. Ethical philosophy is quite different from this and often entails what Derrida has called the ordeal of the undecidable ("Force of Law" 24). Ethical philosophy holds that there may be no stable parameters or prescriptive rules by which to judge ethical and unethical actions. The decisions one makes in the wake of such uncertain conditions involve understanding as much as possible and carefully thinking through matters that almost always involve relations of power and inequality and taking full responsibility for those decisions. Ethical philosophy is a rigorous mode of thought that is not inclined to produce easy decisions or clean consciences. The question of what exactly is involved in ethics is not a settled matter; thus, setting out bioethical rules to minimize suffering for animal test subjects, for example, is a decision that may involve already deciding what kinds of experiences will be recognized as suffering. Or, one might think of the use of animals whose entire life will be in a laboratory governed by rules for proper treatment and ask: Is this not already an incredible violence that cannot be undone by any effort to prevent mistreatment?

Bioethics thus begins within a framework and a set of assumptions that it may not regularly notice, let alone examine, including certainty regarding the categories of the human and nonhuman animal. As Wolfe and others have shown, the desire to justify such "treatment" of nonhuman animals comes from a speciesism that sees animals as something that humans can treat as they wish, so long as we follow a particular set of rules decided by humans as to what is acceptable. Bioethics tends to decide in advance on the necessity of a course of action and then decides under what circumstances this action will be permissible. In the harsh light of our technical capability, activities are sometimes justified because we can conduct them rather than ask truly challenging questions about the assumptions that subtend those activities, such as what respect we owe to nonhuman animal life on this planet and how we have come to view it as our right to decide its fate.

The logic is similar for biometric practices in the sense that it frequently involves a pragmatic approach of thinking about best practices. In instances of racial profiling at the airport during security screening, voicing disappointment about individual and state forms of prejudice can produce pragmatic improvements such as better training to ensure that a properly neutral technique of checking identities does not serve divisive, racist purposes and discriminate among travelers. Yet a pragmatic contention that the technology is only as good or bad as the people who implement it prevents us from understanding the ways in which biometric thought naturalizes a range of ideas, including the ideas that human movement ought to be governed and that openness and transparency are a moral good that can be determined within only particular parameters. Can we say that a technology that insists on limiting a person's identity to their body is neutral? For individuals who feel at odds with their bodies in some way, or who experience a sexuality that exists in ways that are not readily manifested by the particular anatomy of their body, this may be an unwelcome presumption about the ways in which the physical body represents oneself.

Can we say that a technology that insists on reducing identity to a singular self is neutral? If we are social creatures whose identities depend upon social norms of appearance and well-developed social conventions that make up something called identity, and if it is on the basis of these social concepts that we become legible to ourselves and others, then the power of a biometric portrait of existence to strip away all of that profoundly alters how we think about human existence.

Can we say that a technology that is unable and unwilling to comprehend context and history as features of identity is neutral? Without anything but the present individual, one is left with a form of identity that cannot recognize how one arrived in the present or how one might live in ways that are just and meaningful and able to record and redress histories of persecution and injustice. Before biometrics scans an iris, analyzes a palm, or recognizes a face, it has made a number of decisions about what is and is not part of a practice of identification and official recognition.

Biometrics can really only claim neutrality if we forget the ways in which it has already defined a field of applicable and non-applicable concerns. To the extent that biometric access control can involve some of the most consequential matters of human existence—including migration, rights of refuge, and recognition of one's existence by national governments as well as international bodies—how it envisions the person and the kind of borders it places around a concept of personhood means a great deal. Even if biometric thought is far from total in its reach and effects, it nevertheless possesses significant force in defining how we view ourselves as discrete entities largely independent from others and can normalize a strange existence in which we exclude history, culture, and social relations as the basis for understanding ourselves.

Biometric thought, moreover, refers to not simply a set of practices but also a set of desires, fears, ambitions, and concerns that have been given a tangible form. Can the idea of biometric identification at a border *not* raise the specter of terrorism? Can it make one *not* think that immigration is always a matter of discovering the illegal immigrant? Biometric thought creates a view of human movement premised on suspicion. What disappears from view under such a way of thinking? At the very least, what tend to escape the biometric gaze are questions about what engenders migration and what that might have to do with economics and global inequality and the political activities of other nations. Biometrics conjures up an idea of the individual, present at the border and purged of social circumstances and entirely self-possessed in one's motives and desires.

The ideology of neutrality, or the idea that the claim to neutrality is a way of not discussing all the conditions and assumptions that structure biometric thought, can be illustrated in another way. Consider again Magnet's critique of the racism of biometrics. This critique may, perversely, conceal by revealing. By emphasizing potential and actual procedural racism, Magnet focuses upon a set of questions around race that can, as I noted, be addressed and recalibrated so as to eliminate this racism. Such an approach may also, however, make it harder to comprehend and appreciate the ways in which racism, along with other struggles of power and wealth, structures global relations in such a way that certain situations—be they crimes against humanity, impossible economic conditions, political persecution—receive muted responses at best under biometric thought. Seeing the racism that targets specific individuals for added screening may make it more difficult to comprehend another form of racism that justifies the international community's neglect of those suffering in the rebellion and civil war in Syria. By ritualizing the discrete scene of biometric inspection, we may make it more difficult to comprehend and see the larger set of power imbalances that occasion the deployment of biometrics in the first place. There is no reason this must be the case, however. We can and ought to be sensitive to both individual and structural violence, even if we routinely attend to the former and ignore the latter. The following passages, from an article on *Frontline* in December 2015, are fairly typical in how they attend to the technical accomplishments of biometrics in a way that makes the world that requires them disappear:

> In all, more than 4 million refugees have poured out of Iraq and Syria due to the fighting in each country. Because many refugees flee their homes without papers, or lose documentation in their shuffle through borders and camps, they can be difficult to identify and track. Until recently, fingerprinting was the United Nations' favored identification technique.
>
> That changed in October 2013, when the U.N.'s refugee agency adopted iris scanners. Other than very young children, whose eyes are still evolving, and people with certain kinds of eye damage, any individual can be identified in this way with tremendous accuracy and speed.
>
> In the last two years, the agency has scanned the eyeballs of more than 1.6 million refugees in nations across the Middle East and Europe, with the notable exception of Turkey, which still insists on fingerprints, said Larry Yungk, a senior resettlement officer with the U.N. The U.N. now has a "fairly complete biometric database" cataloging the iris patterns and identity of the Syrians and Iraqis who have fled their homelands, Yungk said.
>
> Using the scanners, the agency can track where and when refugees check into camps and offices. It also deters fraud, confirming the identity of applicants for aid, services or relocation. In some countries, the U.N. has even made deals with banks to attach iris scanners to ATMs, so that only those authorized for assistance can make withdrawals. (Worth)

Deutsche Welle similarly reported on measures to register the more than one million refugees who arrived in Germany in 2015:

> Germany's federal parliament on Thursday approved a plan that will provide identity cards linked to a centralized data system to refugees.
>
> The new law is aimed at providing authorities a way of keeping track of those entering the EU's most populous country after some 1.1 million refugees arrived in 2015.
>
> The legislation was previously approved by German Chancellor Angela Merkel's cabinet.
>
> Beginning in February, refugees registered in Germany would receive one ID card containing all the information required for an asylum request, according to officials.

The IDs will include information such as fingerprints, country of origin, contact details, health status and qualifications.

The new system is expected to be fully implemented by the summer, allowing all government agencies access to the centralized system.

The move comes after criticism of Germany's decentralized system, which allowed some migrants the ability to fake their identity or register multiple times. ("German")

In each instance, biometric screening is hailed for its efficiency in the face of a substantial scale of refuge seekers and the global crisis they represent. Yet what is entirely absent from each account is attention to the actions that led to such a crisis. *Frontline* mentions fighting within Iraq and Syria but does not identify how the U.S. invasion and occupation of Iraq since 2001 led to conditions in which ISIS (Islamic State of Iraq and Syria) has flourished or the international community's muted response to war crimes in Syria or its failure to protect civilians in that country's civil war. The turn to the technological is a mode of not thinking about what has led to this crisis.

One might object and say that this is a category mistake and I am expecting technology to address world issues. I think this is precisely what biometrics is seen to do in these accounts, however. It is a technological solution to a problem that reduces what is visible and legible about that problem to a very narrow set of concerns. Such accounts focus on individuals and identity and integration as well as support for refugees, and these are all ways of thinking and speaking about a set of circumstances without discussing responsibility or history or the causes and conditions of migration.

Even comparing these instances to a far more sensationalist article in the tabloid the *Daily Mail* reveals a similarly selective attention that is guided by biometric thought, noting that the absence of advanced biometric technology makes Syrian documents less secure:

Authorities in Germany are powerless to halt the asylum claims of tens of thousands of migrants using fake Syrian documents, MailOnline can reveal.

Officials registering applications have been overwhelmed by the 500,000-plus refugees who have streamed into the country since the beginning of the year.

A huge number of claims for asylum that are supported by passports and identity cards that appear to be false, a police forensic expert has revealed. [. . .]

Police forgery expert Joerg Aehnlich told MailOnline: "I know their documents are false but I cannot give evidence in court that they [the asylum seekers] are not Syrian because I cannot prove it." Migrants are using documents stolen from Syrian refugees, identity cards manufactured to order or simply papers borrowed from friends and relatives to support their asylum application, Mr Aehnlich, of the Lower Saxony Criminal Forensic Institute has revealed. [. . .]

Germany expects at least 800,000 asylum seekers this year, with some estimates as high as 1.5 million. Some 15,000 migrants crossed into the country from Austria in a single weekend. [...]

In Turkey, which hosts up to 2.5 million refugees, Syrians need a passport to rent a flat, stay in a hotel, open a bank account and even to buy a sim card.

Forgers buy "empty" passports without pictures or information from government officials, or from fighters who seize them from government offices.

Syrian passports do not have modern security features such as biometric chips, making them easier to forge. [...]

The Syrian Government announced this month that it has made more than $500 million in passport fees as it doubled the cost of buying a legitimate Syrian passport abroad from $200 to $400 for a brand-new document, and $200 to renew. According to Al-Watan, a pro-Assad daily newspaper, the Syrian government issued 829,000 passports this year - about 3,000 a day. (Fagge)

It is not surprising to see such scurrilous reporting from a tabloid source, and its characteristic dehumanization of those seeking a livable life has become unimaginably routine in so many more legitimate sectors of news media. What is especially noteworthy here is simply that the focus on biometrics is a means of sharpening attention on security, criminality, and xenophobia and is an exercise in unveiling the truth beneath the surface: documents are seen to be more vulnerable than people. Under the glare of such biometric thinking, a humanitarian crisis disappears, as does a crisis spurred on by wealthy nations such as the United States and Britain, which created and sustained the geopolitical conditions that made Syria into a conflict zone and allowed it to remain unresolved and its citizens vulnerable. What appears instead is a scene of inspection in which those who arrive in Europe are identified and found to be suspect; what follows is an opportunity to affirm hatred and justify, in advance, for British readers, the exclusion of those desperate to survive in the wake of war.

Biometric thought is not neutral in the sense that its effects cannot be limited only to discrete moments of identity authentication. Even when authentication may be all that is at stake, biometrics still depends upon a particular mode of excising the individual from the world that sustains existence. Against claims that biometrics is simply a technique and a tool, I insist that it is imbricated in a social process of determining what counts as truth and how that can be recognized as such. Biometric thought especially constrains how we think about a number of pressing social matters that arise thanks to human migration and the social and economic abandonment that makes migration necessary.

Biometric thought can condition how one thinks about particular events and police the lives of those directly and indirectly affected. These operations are shaped by social practices and assumptions, and they seek to intervene in the world to shape it, define its possibilities and potential futures—not just represent it. It is a mode of knowing that acts in the world. This supposition that a way of seeing reality is also a way of shaping what we can know and conceive of can be traced back to Immanuel Kant's understanding that the emergence of rationality and reason over the course of the eighteenth century was a human social development that was also a way of ordering that reality and making it comprehensible by emphasizing some details and ignoring others. When Kant asked the question "What is Enlightenment?" in a Prussian newspaper in 1784, he set out to characterize an era of philosophical thought but also what the Enlightenment and rationality do in the world. It was a philosophical inquiry into what Foucault would later call "the nature of today" ("Structuralism" 443), which begins with the recognition that a mode of thought is less a representation of reality than a way of forming something out of social existence that can be recognized as reality.

What is biometric thought and how does it mediate the nature of today? What is the nature of our biometric present? To ask this question means recognizing that biometric thought is continually modified, put to work, and made useful and that it makes intelligible an entire natural, social, political, and cultural reality that intersects with a range of power relations and continually charged desires that can reflect and renovate norms and social ideals. To speak of biometric thought is to address something older than the present and which organizes the present in particular ways. It is not true that the proper or natural order of things "was subsequently diverted by such-and-such an event"; instead, what we think of as the present is, in part, the product of ways of comprehending it and making it legible (443). The present is not inevitable, then; it could entail "different foundations, different creations, different modifications in which rationalities engender one another, oppose and pursue one another" (443). The present we have is an effect of social modes of knowing that condition reality and what we can imagine about it. And technology plays a significant part in how we come to understand the world. Technology does not arrive from another world. It cannot be excised from the history of modes of understanding human existence. Technology is always a human technology built upon human desires, fears, anxieties, irrationalities, presumptions, dreams, and nightmares. I follow Foucault's analysis of the productive force of knowledge by asking how biometric technology both reflects and conditions particular ways of understanding social and political realities and how its operations, ambitions, and uses have changed over time.

If the question that structured much of this chapter was "What is biometric thought?" we now see it may be even more important to ask, "What does biometric thought do?" What kind of person, for example, does it imagine will arrive at the border or slip past unnoticed? Individuals are not vulnerable, resourceful, lively, dedicated survivors under biometric thought. They are unknown threats. They are potential cheats and scoundrels and criminals looking to take advantage of the generosity of others. What kind of advancement is implied by this technology? The idea that biometrics is desirable because it is an efficient technology suitable for authenticating individuals quickly may itself be a dangerous assumption if it reduces human movement to a very particular solvable problem. The prospect of efficiently sorting the wanted from the unwanted—whatever that means!—or regular migrants from irregular migrants normalizes human movement as something to be regulated and managed instead of a matter that prompts another question: What is it about the way in which individuals have been economically and socially abandoned that leads them to migrate to other places in order to survive? If stalling lives within refugee camps appears unexceptional, or if humiliating a welfare claimant by insisting she would, given the chance, defraud the government, if this all becomes accepted as a reasonable way of seeing the world, then we produce forms of slow death and neglect variously in the name of protecting the interests of the nation and policing human mobility. This is not something that simply happens. This is something we make possible by choosing to accept the designs and ambitions of biometric thought.

The idea that biometric thought produces particular effects and generates particular fields of truth can be approached in other ways as well. The rhetoric of biometric precision supplements a glaring lack of precision in determining who and what might be a threat to a given society. Could it be that technical procedures of identification stand in, reassuringly, for the vague use to which it is put? This might explain the breathless technological appreciation that often features in assessments of the science. Such automation and certainty contrasts powerfully with the poorly defined terrorist threats that border patrols hope to pinpoint and which serve to justify ever-expanding budgets. In an interview, Groebner highlights a critique of procedural biometrics that takes aim at exactly this air of precision associated with biometric thought:

I'm sorry to say that, from a historical point of view, the new field of biometrics— the gathering of behavioral or physiological data about an individual—may well follow the way of its forerunners. It will be doomed because it will pile up so much information about the actual encoding, registering, and collecting apparatuses that the biometric data itself will become secondary. The process of collecting and ordering huge masses of data is more energy consuming than those who build the registration apparatuses realize. The accumulation of biometric data will fall victim to the same entropy as the huge piles of paperwork in the archives of King Philip of Spain who, at the end of the sixteenth century, wanted positively to establish the identities of everyone emigrating to the New World in order to prevent the offspring of Jews, Arabs, convicted heretics, runaway priests, and debtors from wandering across the Atlantic to start a new life. But that is what they did. All data run the risk of turning into a bureaucratic fantasy in which the already stored information is used to provide the categories of authenticity for new information. (Serlin)

A regime of truth emerges from such practices of data collection. A biometric practice of identification defines what will be assessed, what will not be examined, and what a given category can say about a person. In addition to subjecting reality to an organizing principle that makes some matters obvious and others almost inconceivable, a regime of truth also makes possible a number of quite consequential effects. Regardless of the data collected and its usefulness, the insistence on a biometric logic of access control regulates how individuals move about the globe and constrains how we understand that movement. Biometric thought conditions how individuals are asked to understand themselves and the basis of their existence. These regimes of truth are not the only ones possible, however. They exist because they serve particular ends or represent particular social anxieties. The dream of biometric precision forgets the depth of human dignity and the manner in which identity depends upon a social world and the context that lies beyond an individual person. Its truth, focused as it is on the identification of individuals by bringing tangible unique features into view, may forget the economic and historical realities that organize how individuals and communities exist or the global relations of power that structure so much human movement.

The desire for biometric control at the border references the extent to which many states cannot control the transnational economic conditions that can drain and enrich individuals. The movement of investment capital and finance can abandon people and spur individuals around the world, impoverish them as well as create opportunities, construct communities, and deplete the earth for the enrichment of a few. What happens in one part of the world is, more than ever before, directly linked to what goes on in another, whether this is because of resource extraction or the manufacture of goods. Biometric control of human movement is one public face of an economic reality in which states cannot or will not regulate financial capital and its effects upon the lives of so many within their territories. It may be nothing less than a direct means of regulating the worst effects of capital for an age when investment and profit move effortlessly around the globe and with minimal regulation. While I am not convinced that economics can fully account for all the operations of biometrics, it is clear that biometrics has a social life that belongs to a much wider sphere of public life, including a global economy that is responsible for so many of the conditions that shape how individuals live, whether or not they can thrive, and whether or not their dignity is recognized as equal to the dignity of another elsewhere.

Perhaps it is surprising, given my interest in the social life of biometrics, that I dwell only infrequently on privacy, so far. The capacity to collect and expose important data or to share it in ways that violate the terms within which it was originally collected can be of tremendous consequence. But my focus is not on how information is stored or the ways in which it might be made public, though I will address how biometric thought is concerned with privacy and a view of the body as a private reserve of oneself apart from the world. This notion of the body is key to a history of biometric thought, and it has a history itself, including social norms regarding what is private or public about oneself. As opposed to an idea that the body ought to remain untouched by technology, I find myself wondering whether the body is itself a technology, in the sense that it is part of both me and a social schema that has been formed by sedimented perceptions and assumptions.

I am reminded of a beautiful commentary by Derrida on the technological supports that make writing and thinking possible, supports that philosophy likes to forget as much as it likes to forget that the body is a crucial part of our existence: "as if that liturgy for a single hand was required, as if that figure of the human body gathered up, bent over, applying, and stretching itself toward an inked point were as necessary to the ritual of a thinking engraving as the white surface of the paper subjectile on the table as support" (*Paper Machine* 20). This does not imply that the body is a thing like any other thing but rather that "thinking engraving"—what a wonderful phrase for thought and the way it etches upon ourselves and others—involves a physical existence in the world that mediates and shapes how we live and think in the world. Our ideas do not exist outside of the world of lived relations and the possibility of engraving them, recording them, and sharing them with others, and this is part of what I hope to respect by speaking of biometric thought as a mediating force that shapes how we think, live, and know ourselves and others.

My emphasis on the idea of regimes of truth engraved by biometric thought ought to be distinguished from an established practice of critical thought that is premised on a logic of concealment and unveiling that sometimes affirms the "methodological centrality of suspicion to current critical practice" (Sedgwick, *Touching* 125). Such critical practice contends that reality has been distorted and believes that if we see what is truly there, we will be able to produce a better future on the basis of such open knowledge. It sounds absurd to disagree with this premise and argue for *not* seeing or, worse, being in favor of a world one knows to be premised on lies. While it is true that some violations of human decency can be stopped if they are brought to light, I am wary of an approach that suggests that (1) shining the light of day is all it takes to reverse forms of dehumanization and (2) if we are to be critical of the injuries that are perpetrated against so many, we must be as paranoid as possible. Suspicion stalls thinking by offering up simple identifications of powerful individuals and interests who control the rest of us. Worse, such paranoia may be debilitating: How can I ever know if I have it right and have arrived at the bottom of things, at the real truth of what is going on? Should I wait to act until I am sure? Paranoia means never being paranoid enough.

Instead of trying to see what is *really* there, my turn to biometric thought is an attempt to see what makes sight possible and acknowledge some of the effects of a given mode of recognizing and making sense of reality. This is no longer a biometric process of unveiling and discerning the truth below the surface but a matter of asking after the conditions that produce the knowledge we have and do not have. Because biometric thought involves more than a set of procedures for verifying identity and more than a set of security measures, it can be found interacting with many aspects of social and cultural life.

In the next chapter, I want to highlight some of the central domains of biometric thought. These domains do not identify all the effects of biometric thought, nor do they specify what it means to live a life made more difficult by its operations or to exist as if without a world of social relations. These domains exist as cultural concepts and practices influencing and affected by biometric regulation. Rather than thinking of biometrics as a technical measure handled by border guards calling out "Who's there?" as we saw in the introductory discussion of *Hamlet*, these domains show that biometric thought is attached to an existing set of ideas and assumptions and that these form the basis for how it orders the world and the regimes of truth that it produces.

3 · THE DOMAINS OF BIOMETRIC THOUGHT

If one has ever felt anxiety at a moment of inspection, perhaps at a border, it can invoke an odd sort of trepidation when realizing the impossibility of acting in a manner consonant with whatever the official record of identity might suggest or in a way that seems utterly "normal." The ritual of identification can be like getting thrust on stage without a script, wondering how to perform a calmer, more lively, upstanding, relaxed, happy, serious, obsequious, or assertive version of oneself. It is as if the official record, itself a fiction of identity that has reduced the individual to a series of details deemed relevant, becomes the ideal that one must reproduce. More, can one be sure how this performance will be judged, what details about oneself will be noticed, and what conclusions will be drawn from them? Identity becomes something beyond oneself, despite its intimacy, because it is up to another to confirm.

Some may contend that biometrics is limited to exceptional experiences of identification largely reserved for boarding airplanes and crossing borders. And, as something unusual, it is of little significance for how we actually live in the world. For some, it amounts to little more than an expectation grudgingly met and is, for most, a tolerable inconvenience. One might reason, "If it distorts who I am by transforming me into binary code, so be it. This will not transform who I am when I arrive at my destination. And even if such scrutiny is uncomfortable or unwelcome, I can tolerate it in the name of safety and security. It is neither too invasive nor too unreasonable, if executed appropriately."

What accounts for this eager desire to dismiss the consequence of biometric inspection, even when acknowledging that it touches upon matters as significant as identity and its relationship to the physical body, where and how one lives, and national security? What knowledge are we keeping from ourselves by seeing biometrics as something of little consequence? Why are we so ready to admit the anxiety that facing authority can produce, yet so eager to dismiss those feelings? Is there really only one representative experience of what it means to be subject to biometric identification?

Dismissing the significance of biometrics means isolating its activities and ignoring the ways in which its logic supports and is supported by many other social norms and ideas that enable its operations. Seeing biometrics as only a practice of momentary inspection affirms that such episodes are isolated from the so-called real world, and it subtly implies that outside of that moment in which identity must be verified, one is otherwise in control of the terms by which one exists. Indeed, a certain sense of one's independence may depend on dismissing a moment in which identity can only be officially confirmed. Dismissing biometric identification makes it possible to forget that each of us exists in ways we do not absolutely control or choose. By seeing biometric inspection as a relatively inconsequential moment, one likewise insists that everyone experiences biometric inspection in the same way. What ought to be immediately apparent is that biometric inspection, or even just the possibility of it, is a great deal more consequential for some than it is for others. To insist that it is inconsequential, then, voices a normative claim that everyone experiences this reality in the same way and that every individual possesses the same capacity to claim an identity that can receive permission to move about the globe or have one's identity verified because it is stable and because effective records exist to document it. Biometric thought tends to see all individuals as fundamentally the same in their capacity to be identified and fails to acknowledge the very different circumstances that shape them and how they live.

The experiences of refugees, the economically abandoned, persecuted populations, indigenous peoples, and undocumented migrants all attest to the ways in which biometric identification may be experienced in profoundly consequential ways—and not just because it is an experience that can make life more difficult or regulate where and how someone can move on the planet. For some, biometric identification may helpfully supplement withheld or unavailable identity documents, just as others may find themselves criminalized for not participating in official rituals of identification. My point is twofold: First, that what biometrics does and how it affects someone is going to differ in individual situations; thus, the claim that it is insignificant is a normative one that refuses to acknowledge the different power relations of inequality and injustice that structure social existence. Second, that dismissing biometrics as inconsequential dismisses the differences between lived experience on this planet and simultaneously dismisses the possibility of ever recognizing and thinking through the significance of those differences.

It is not just travelers who perceive biometrics in ways that tend to isolate its reach and acknowledge only some of its effects. Indeed, some of the scholars interested in biometrics that I have considered thus far see in biometrics a decisive confirmation of an omnipotent state that monitors our every move, and this articulates a position that is actually quite close to the dismissive hypothesis. It strikes me that the desire to say that biometric identification does not matter at all and the desire to say that biometrics confirms our worst fears about authoritarian surveillance share a similar refusal to acknowledge the complex social life of biometric thought that regulates how we think about ourselves, others, and the world around us. Seeing biometrics as extraordinary, each of these perspectives fails to comprehend how biometric thought is powerfully attached to many of the primary conceptual and normative domains of social life.

As a mode of thought, biometrics has a social life attached to cultural norms and ideals that condition concepts of security, migration, identity, the physical basis for existence, and the border, among others. It is important to note from the outset that none of the domains I consider in this chapter are entirely determined by biometric thought. Each are quickened in different ways by biometric thought. For example, how does biometric thought reshape what it means to have an identity that can be stolen? Or, how has the insistence that identity is attached to the physical body made other ways of thinking about oneself less viable?

I hope to demonstrate the power of biometric thought to transform how we think and note how biometric thought is attached to and preoccupied by particular concepts and concerns. Examining these domains makes it possible to see how biometric thought builds upon existing features of social and cultural life and how they come to produce the effects they do. If biometric thought depends upon particular conceptual domains for some of its primary social effects, it does so by altering relations of power or inscribing them in new ways or toward new outcomes. What also becomes clear from this examination is that the regulatory force of biometric thought does not issue from a single source, such as an all-knowing agency conducting surveillance, but instead grows out of particular interactions, measures, procedures, and ideals attached to particular domains. It is a normative mode of thought as well as a direct means of regulation, and it issues from multiple sources and in inconsistent ways. What follows is not an exhaustive account of the attachments that structure biometric thought, then. Rather, it is an attempt to mark and describe some of its most powerful and consequential domains. What follows after that are two extended illustrations of how these domains intersect in particular instances of biometric thought.

IDENTITY

Biometrics can create an identity that is immutable and permanent but which is likewise beyond one's own capacity to verify. This does not mean I know nothing about my identity but that it requires the intervention of something or someone beyond me. My identity belongs to me to the extent that it is seen or recorded and verified by others. This might involve advanced technological measures or simple official documents. I suspect this may be what is popularly unwelcome about biometrics: it feels alienating because it takes away my power to give an account of who I am. I can only read from the script that has been authorized as a legitimate way of speaking about myself. I can no longer say what I want and must instead let my features and my recorded history speak for me.

Another way of putting this would be to say that biometrics determines which parts of me have the power to narrate who I am and under what circumstances they can do so. These parts may or may not be features that I tend to think of as decisive indicators and expressions of my identity. Thus, biometric thought powerfully reveals how we live with and represent our identities and come to recognize that whatever else an identity entails, it represents a social existence that is partially beyond my control and reach because I do not determine the terms within which my identity exists or becomes a functional concept. Whether those features are familiar or strange, such as one's voice or one's fingerprint, they are not matters that I select or that I alone imbue with a particular force of representation. They are social categories that are understood to be markers of identity, regardless of what I might wish.

As I noted earlier, "Who's there?" is both a question and demand to submit to identification. For biometric thought the two operations are inseparable because there are only certain types of responses that will be heard in reply to this question. Identification is a matter of submitting to procedures that have the authority to authenticate and produce truth. I began this chapter by examining the claim that our biometric identity is the one we do not choose, as opposed to the one we do choose. It is, according to this logic, an ill-fitting uniform into which I must briefly squeeze myself if I am to respect regimes of identity authentication. But this is a false distinction because our "real" selves are not of our choosing, either. They likewise rely upon a constrained field within which certain responses will be expected and acceptable, while others will be unacceptable according to the social norms that structure how we represent and interact with identity. I cannot, for example, opt out of norms of sex and gender and simply refuse the social codes of legibility they bring to bear upon me in my society. I might choose to try, but others will still use them to read me and grasp the nature of who I am. As the philosopher Judith Butler explores, how we experience identity means more than a desire to identify ourselves. It means fitting oneself into forms of narration that pre-exist us and which signal the ways in which identity involves social recognitions and norms of intelligibility:

> If my face is readable at all, it becomes so only by entering into a visual frame that conditions its readability. If some can "read" me when others cannot, is it only because those who can read me have internal talents that others lack? Or is it that a certain practice of reading becomes possible in relation to certain frames and images that over time produce what we call "capacity"? For instance, if one is to respond ethically to a human face, there must first be a frame for the human, one that can include any number of variations as ready instances. But given how contested the visual representation of the "human" is, it would appear that our capacity

to respond to a face as a human face is conditioned and mediated by frames of reference that are variably humanizing and dehumanizing. (*Giving* 29)

Reading the face of another depends upon a set of learned associations regarding particular features and what they suggest. We are always operating within some reference to an idealized social norm that polices how we grasp the world and those we encounter within it, and these are not norms we have created or consciously chosen.

Identification is governed by modes of thought and norms that I do not control or determine but which pre-exist who I am. If we understand that identity always involves some degree of dispossession, some degree of not me at the core of who I am, this may indicate part of what is unsettling about the experience of biometric inspection: it means coming face-to-face with my own strangeness and the ways in which I rely on matters not my own to comprehend myself. This does not mean that I cannot be the sort of person I want to be, necessarily, only that I risk social illegibility if I opt to fashion myself in profoundly unconventional ways and that there may be social ramifications that are more or less easy to bear for doing so.

Biometric thought fixes identity to the body and, at the same time, to social norms that condition its legibility. This means that identity emerges from establishing accepted way of seeing individuals, such that biometric verification cannot, for example, recognize the importance of the land or the significant ways in which we are bonded to family or a culture or language or the power of history in shaping who one is. Identity becomes something embodied and discretely individual, even if the individual cannot verify it alone. More, biometrics suggests that identity is not something we develop over time but something that exists almost outside of time as a permanent feature of our embodied lives. Indeed, part of its promise is that identity is not affected by time or desire: a biometric profile remains constant and verifiable over time, even and especially if I do not. Biometrics speaks in place of a narrative of oneself and the opportunities one routinely recognizes as one's own to fashion oneself and indeed to grapple with the conditions of one's emergence as an identifiable and singular individual.

What I would like to suggest, further, is that for biometric thought, identity emerges as a site of conflict rather than something obviously and plainly given. Biometric thought helps to produce conditions that estrange us from the obviousness of identity; it is predicated on a notion of identity that is social, involved with normative conventions that govern how and to what extent individuals are legible, even to ourselves. To be sure, the idea that identity is a site of conflict— continually being worked out, contested, revised, insisted upon, bargained with, given up, lost, and remembered—does not seem to be a desired outcome of biometric thought, but this is, nonetheless, a potential effect of the alienating experience of biometric inspection.

PRESENCE

Biometrics presents the possibility of attaching identity verification firmly to the body, as travelers at Rajiv Gandhi International Airport in Hyderabad know well. Individuals check in online and then submit a fingerprint scan at the airport in order to verify their identity if they have previously registered for an Aadhaar Identification Card. Such a paperless world is presented as a new form of efficiency and convenience for travelers, made possible thanks to biometric technology. All the traveler needs to do is present oneself as an expression of identity (Lee). Biometric practices are presented as more secure and more real by virtue of the way they grab hold of the physical body as the ground for identity. Biometric thought trades on a sense of the immediacy and presence of the body, even if it ultimately transforms this presence into mediated representations of the individual that are not much different than older, less technologically complex paper records of identity.

Biometrics intensifies a concept of presence and significantly shapes how we think about presence and what it entails, especially the presence of individuals in public. Consider, for example, the concerns surrounding the idea of presence governed by a law in Arizona that both authorizes and compels police, on the basis of reasonable suspicion, to determine an individual's citizenship and immigration status. This law, known as SB-1070, mandates racial profiling and creates conditions that make the public presence of some a persistent criminal offense and effectively bans undocumented individuals from accessing police protections. By contrast, Los Angeles has had ordinances in place since 1979 that prevent local government officials, including police, from asking about immigration status as a condition of providing services.

Arizona's biometric impulse to identify and document faces on the street—and let's not pretend that this is a law designed to do anything less than legally codify the Latino community as the general face of criminality in Arizona by making them subject to persistent state aggression and racism—creates situations in which simply being present and visible in public makes one subject to a test of immigration status. If one can be forced to identify oneself on the basis of biometric suspicion, one's existence in public is constrained by a calculus of risk, and one's existence, dignity, and political presence are reduced to identifiable markers such as skin color or language spoken. One is forced to live a life governed by the impossibility of ever being recognized to have any right to exist in this territory, regardless of nationality. If the body can be transformed into a form of documentation, it means that all moments of public presence become opportunities for inspection, such that the logic of the border now becomes a permanent and selectively applied logic that governs all of public life. I want to be clear: this law does not mean only that some are subject to ID checks. This law is an attempt to negate the public sphere itself such that the appearance of anyone who appears

Latino is understood in advance to be a suspect, not a human being possessed of dignity. If individuals cannot appear in public, the very prospect of the public is jeopardized.

Such a law, moreover, criminalizes indigenous rights to access territory that spans either side of the border and empowers police to treat as criminal indigenous populations living on tribal lands if they do not carry identification. The logic of presence overwhelms all other considerations and rights, and daily life becomes a limitless reminder that one exists on the land at the state's prerogative and that rights to territory and sovereignty are trumped by the insistence that some are less than human, rightfully dispossessed of security, excised from an enduring relationship to the land, and less able to make nation-to-nation claims regarding territory that straddles the border separating Mexico from the United States. Biometric thought sees borders as checkpoints or walls that must curtail movement, and this logic is especially objectionable for the Tohono O'odham, an indigenous people who have always lived on territory that now spans seventy-five miles of this border and a total of 2.8 million acres (Kilpatrick). Individuals on this land are arbitrarily distinguished by the state's exclusive recognition of what kind of documentary evidence of identity is legitimate, such as citizenship records, and what is not, such as residence, culture, traditional knowledge, and language. For this reason, they have long opposed any form of border wall:

> The Tohono O'odham have resided in what is now southern and central Arizona and northern Mexico since time immemorial. The Gadsden Purchase of 1853 divided the Tohono O'odham's traditional lands and separated their communities. Today, the Nation's reservation includes 62 miles of international border. The Nation is a federally recognized tribe of 34,000 members, including more than 2,000 residing in Mexico. Long before there was a border, tribal members traveled back and forth to visit family, participate in cultural and religious events, and many other practices. For these reasons and many others, the Nation has opposed fortified walls on the border for many years. ("No Wall")

Biometric practices might focus on individual bodies, but their application has the effect of transforming territory and culture for the Tohono O'odham. When such operations of controlling access intervene legally to decide upon contested notions of who people are and where they belong, they alter everyday life by forgetting that there is more than one way to understand a territory and those present upon it. *The New York Times* quotes one resident who notes "there is no O'odham word for wall," and perhaps equally telling, the article explains, "there is also no word for 'citizenship'" (Santos). How someone is recognized to be present and what sorts of ideas an individual's presence conjures are rarely interrogated by a biometric thought that sees the human as a mobile unit to be identified as belonging to one place or another according to those who create maps and build walls. As the Tohono O'odham show with their compassion toward migrants, this view need not be exclusive and can admit other ways of seeing a particular territory and those present within it:

> Tohono O'odham leaders acknowledged that they were straddling a bona fide national security concern. The tribe reluctantly complied when the federal government moved to replace an old barbed-wire fence with sturdier barriers that were designed to stop vehicles ferrying drugs from Mexico. It ceded five acres so the Border Patrol could build a base with dormitories for its agents and space to temporarily detain migrants. It has worked with the Border Patrol; hardly a day goes by without a resident or tribal police officer calling in a smuggler spotted going by or a migrant in distress, said Mr. Saunders, the director of public safety. The tribe regularly treats sick migrants at its hospital and paid $2,500 on average for the autopsies of bodies of migrants found dead on its land, mostly from dehydration. (There were 85 last year, Mr. Saunders said.) (Santos)

BORDERS

Borders were, for a long time, best viewed on maps. While border checkpoints exist, the actual physical demarcations of the line where one country stopped and another began were not especially visible, except when it was marked by physical terrain such as a mountain or a body of water. This may be less true now than ever before in history, with the rise of walls around so many states. Yet walls are, as Brown so perceptively notes, physical expressions of "unsettled and unsecured sovereignty" (90). They are giant monuments meant to reassure all who see them of the integrity of national borders effectively regulated by the state. And such proof is all the more necessary considering how tenuous the claim to sovereignty has become in an age in which "transnational flows of capital, people, ideas, goods, violence, and political and religious fealty . . . tear at the borders they cross and crystallize as powers within them, thus compromising sovereignty from its edges and its interior" (22). At a time when neoliberalism teaches absolute adherence to free-market principles over and against legal or political or ethical considerations, the state has never looked less sure of its territory and ability to make its borders matter. While lacking the stature of a wall, a discourse of biometrics similarly helps to reassure populations that borders remain—and remain powerful zones of regulation that can and do control access to a territory. That "border fortifications can have little or no effect on the most dangerous instruments of terror" including "biological and nuclear weapons or hijacked airplanes" (69) are matters, thankfully, forgotten by the technological sophistication of biometrics.

How does biometrics depend on a concept of the border, and in what ways does it conceive of the border? It might be said that what I discuss in this book has less to do with biometrics and much more to do with human movement and borders policed by biometric technology. To the extent that biometric thought very often involves granting access based on verifying identity and assessing the legitimacy of the request for access, I contend that biometric thought is almost always engaged in policing movement via authentication: who can go where, even if that may mean digital rather than physical movement. I emphasize human migration and movement in relation to biometrics because this is among the most public face of biometrics and the one most often cited as its *raison d'être*.

Access is what biometrics promises to regulate, and by *regulate* I do not mean forbid or enable, as if we might reduce biometric thought to a gate that opens or shuts for a given visitor. Rather, *regulate* refers to a process of making movement legible in particular terms that insist movement is a matter of identification first and foremost as well as an object of concern that must be managed. Indeed, the realities of what a border is have come to be mediated by biometric data and identity verification. For the paperless, the border may loom as a barrier that can only be navigated unofficially. For those enrolled in additional voluntary channels of biometric identity verification, the border is an almost vanishing experience. As Ajana notes in her consideration of "advanced border technologies," biometrics has created "the 'kinetic elite', a category of mobile actors who are endowed with private mobility rights and expedited border-crossing entitlements, which exempt them from waiting in busy check-in queues or undergoing lengthy security procedures" (124). The border is not simply a barrier that swings open for some and not for others, then. The border is a concept as much as a location and a set of experiences that are lived differently for different individuals, rather than a stable essence.

The border is a relay in relations of power and a means of mapping and making visible some international crises as problems while placing others beyond the realm of consideration. For example, Canadian discussions of Syrian refugees in 2016 frequently mentioned biometric inspection of individuals and showed scenes of families clearing customs and immigration. And how rarely did these same discussions note that the nation's military was dropping bombs throughout Syria. How infrequently did such discussions even hint at the longer narrative of Western nations variously destabilizing and propping up governments in the Middle East or begin to articulate a discussion of responsibility and the obligation to do so much better for those who managed not to drown in the Mediterranean while fleeing war, rather than imprison them in camps. To insist on biometric spectacles of inspection at the border is to frame this reality in a way that makes a great deal of its complexity and history disappear and makes some of its most important details all but unthinkable.

Conceiving of the border as a barrier importantly misrecognizes how it functions as a conceptual terrain that guides how we think and live. Etienne Balibar notes that the experience of borders is, despite the presence of walls, no longer "at the border, an institutionalized site that could be materialized on the ground and inscribed on the map, where one sovereignty ends and another begins" (220). Elements of economic life appear to exist as if there are no borders at all, while elements of cultural life can be policed aggressively by an insistence upon the border. The effect of the border pervades social life and structures the existence of individuals who will never seek to cross it.

Borders can be vanishingly thin in some instances and insurmountable in others, and not always in ways that are coherent and consistent. For example, trade deals such as the Trans-Pacific Partnership promise to normalize conditions within and between sovereign nations such that the agreement forbids the right of a state to pass laws and institute financial regulations designed to prevent a repeat of the 2008 economic collapse. For Balibar, the very idea of the border has become a space of politics, in the sense that "how borders work" (Vaughn-Williams 8), as well as for whom they work and to what ends they exist, has ceased to be predictable or determined in reference to a consistent set of social or political desires and is instead contingent on shifting concerns and interests and capable of producing intended and unintended consequences.

As a particular way of thinking about the border and its physical presence as a checkpoint and a barrier, biometric thought replaces the considerable political complexity of the border with the simplicity of identification. The greatest simplification may be the way that it represents human movement, as if this was all there was to see at the border. I suspect that the quantities of goods and the movement of financial capital and economic operations in general far outnumber people traversing national borders. Where the border exists as a checkpoint for human beings, it is increasingly experienced as a zone of incomparable freedom for these nonhuman entities. As Ajana notes, "while the free movement of capital, commodities, information and so on is encouraged as it sustains the doctrine of free market and perpetuates global capitalism, the circulation and flow of people, on the other hand, is continuously and vigorously filtered" (55). If biometrics is a figure for national borders—and I would contend it is one of the most common ways we have of representing, as if in shorthand, this occasionally virtual space between nations—it is a figure that withholds almost all consideration of economic deregulation and the lived realities of a global economy that is defined by very selective forms of freedom and movement. Biometrics has become a key part of a narrative that aspires to keep people in place, a narrative which serves the interests of those who profit in a global economy from the presence of financially desperate individuals who are unable to move.

My position entertains the possibility that human movement is legitimate and should be respected rather than prevented or blocked, and this likely strikes some as odd. It is generally true that a nation is responsible, more or less effectively and more or less in practice, for only its citizens. It is not responsible for the citizens of another nation, and thus it ought to be able to decide who accesses its territory and its social supports. This principle, it strikes me, ignores too much the ways in which life within a territory depends upon interactions with life outside of a given territory. For example, the Canadian mining company Goldcorp pays taxes in Canada yet operates mines in Honduras. Similarly, many cheap consumer goods depend on the lack of dignity and respect given to workers in Bangladesh, China, Vietnam, and Latin America. The Syrian refugee crisis that followed from the advent of airstrikes in 2015, moreover, was not an unforeseeable event but the consequence of global failures to act on a humanitarian crisis in Syria and a consequence of the eventual response that tended to be military in scope and which largely added to the destruction of infrastructure and increased the precarity of those trying to survive amid a civil war. Human migration does not happen in a vacuum. There must be some reason for people to walk thousands of miles from their home to an uncertain future and an unknown destination. And it is likely not because one understands oneself to be utterly independent and free from the effects of what others do and fail to do.

World history is, in part, an account of drawing and redrawing borders on the map. What would it mean to write a different history that recorded all the ways in which the borders of the human body have been redrawn? Who signed those treaties, and whose lives have been forever changed by these alliances and pacts? How long has any one set of borders lasted, and what kinds of living continue in the wake of and irrespective of such agreements? As a mode of thought especially attentive to the intersection of psychic and social forces, psychoanalysis continues to record part of that history. Biometrics may well be the concept *par excellence* for thinking about the intersection of political borders and bodily borders and how these affect one another.

THE BODY

Presence is not strictly defined by one's bodily existence in the world; it may have at least as much to do with publicity and visibility, for example. But the idea that we coincide with our physical bodies is a key principle of biometric thought. Indeed, the history of biometric thought consistently links physical existence to official records of identity. Whether that means face reading in the eighteenth century or fingerprinting in the nineteenth or, more recently, passport photographs, ID cards, and the recognition of an individual's unique gait, biometric thought is defined by the articulation of the self in the flesh. There can be no biometric identification without the body, even and especially in digital commerce, where the absence of a face-to-face transaction may be secured via voice-recognition or fingerprint-access controls. The body becomes an expression of oneself and the ground for identification.

Is the body a person? Biometrics can produce interesting interpretations of the physical body as a metaphor of the person, for example, by granting access on the basis of an authenticated fingerprint belonging to a corpse. A physical form may not be the same as the person, even if a person cannot exist without it. What kind of idea of the body does biometric thought produce?

Biometric technology is attached decisively to the body but it has little interest in the world that sustains bodily existence. Individuals made legible by their embodied existence nonetheless exist within a closed loop. The body references the person, who is contained by the body. Nothing else is necessary. This understanding of the body selectively forgets that a body depends upon air, water, food, shelter, and perhaps even a social existence to survive soundly. Biometric thought fixes an ephemeral identity to the body alone, as if that is the only feature that matters for how an individual exists in the world.

One might readily argue that this is only the case for identification purposes, that biometric thought does not prohibit meaningful attachments to the land, for example, even if it forgets they exist. What is difficult to reconcile with this reasonable claim is the way in which biometric thought positions the body as the durable real ground for the person. There is nothing contingent about how biometric thought positions the body. The body is real, the exclusive reality that can verify the person. It does not appear as the most effective or easiest to read; it is the *only* indicator imaginable for the person. So, biometric thought not only reads the body; it makes claims upon the body's exclusive right to be the measure of a person. It accomplishes nothing less than the creation of the most curious specimen: a body that exists as if in a vacuum and solely for the individual of which it is an expression.

What follows from this curious recognition of the body's existence? It reinforces biometric thought's presumption that the body, understood as a ground for the individual, exists durably and beyond degradation. If this body is real, it is strangely unreal, devoid as it is of pleasures and joys or disappointments and aches. It is never "ungovernably desirous and fallible," as Marc Redfield once characterized the body's power to resist being a stable indicator of a person (77). Scholars in disability studies refer to the temporarily able-bodied in order to capture the power of social norms that have led some to believe that a present state of physical or psychic existence is permanent (Marks 18). Both childhood and old age might be thought of as times characterized by alterations in what one can do for oneself. And if one insists on independence or an ideal of looking after oneself as a mark of normalcy, this in itself likely rests on a fundamental misrecognition of the ways in which lives are always supported by others and by social modes that enable and constrain in all sorts of ways. Biometric thought, on the other hand, perceives the body to be blessed with a strange permanence and power beyond degradation, at least to the extent that it is an index of the individual. One can forgive it for dreaming of such a steady state, perhaps, given how jarring it can be when bodies change in an instant or over time.

Biometric thought recognizes the body as the basis for one's independence from others and from the world that sustains it. As a marker of identity, it does not merely represent a present being. It conditions how that being will appear and what can be recognized to matter about that individual. The body comes to exist as a kind of prison that locks one away from others. It is a tactic that shuts one out of other possible affiliations and attachments that might be equally powerful determinations of the essence of one's existence and the possibility of its continuation.

The isolation of the self from others under biometric thought may intensify already challenging social circumstances. Is it surprising that so many immigration detention centers are located on islands, such as those administered by Australia on Manus Island, Nauru, and Christmas Island or by Greece on the Aegean Islands and Lesbos, given the nature of biometric thought? To spatially isolate migrants from the life they have left and from the one they might hope to lead produces a potentially deadly recognition that such individuals are always already unattached to the world around them. Such institutions offer abandonment, isolation, and inhospitality in response to individuals whose presence is in fact a testament to the reality of global interdependence. They fail to recognize that human movement may be largely driven by social, economic, and environmental circumstances and a world defined by their complex integration rather than the isolated desires of lone individuals.

PRIVACY

Privacy, like many of these domains, has a number of different meanings, and not all of them are quickened by biometric thought. Nonetheless, privacy is one of the primary terms within which public commentary on biometrics is mediated. Talk of privacy and biometrics usually addresses at least three quite different concerns. First, advanced biometric technologies promise to guarantee privacy by adding a new layer of security by verifying a person's presence to gain access. Second, advanced biometrics are themselves a violation of physical privacy— invasive in collection and associated with state surveillance—and have led to the creation of government and private-sector databases of unique biological information that may or may not be shared either now or in the future without permission. Third, databases that gather and store information are vulnerable to breaches of privacy.

Much of modern thought regarding privacy begins with "the desire to be left alone," such that we are "free to be ourselves—uninhibited and unconstrained by the prying of others" (Wacks 30). At first glance, biometric thought is not especially good at leaving individuals alone. Biometrics makes claims upon individuals, records their presence, and archives the data accordingly. As a result, modern biometric thought tends to see privacy as a subject of concern and crisis. In a National Research Council book created by the Whither Biometrics group of academics and professionals, the editors note that "virtually every discussion of the social implications of biometrics begins with privacy, and for good reason" (Pato and Millett 100). The book explores the ways in which biometric information can be handled insecurely or shared inappropriately as well as generally expose personal information about individuals to constraints and conditions one cannot control (100). It likewise notes that biometric technologies do "not inevitably threaten privacy" but could be "neutral in that regard or they could even enhance privacy" (100). To appreciate this ambivalence, one need only think about the complexity of biometric data that makes identity theft more difficult by firmly linking the physical body to one's official existence at the same time that it transforms one's existence into information that is more readily reproduced, stolen, or deployed without one's knowledge. Perhaps the most recognizable public face of biometric thought is exactly this concern regarding what happens to stored data that will be used to confirm identity.

Privacy can refer to more than this too and may involve the sorts of scenarios that David Simpson has identified in another context as "lesser evil" narratives (9/11 152): Can a democratic ideal of personal privacy withstand the weight of urgency to save a child abducted from a train station in which video surveillance might be sorted via facial recognition software? Similarly, as a state-mandated requirement for crossing borders legitimately, biometrics tends to create zones where privacy is understood to be a luxury that belongs to another time. Questions surrounding privacy matter. They can and should matter to those who are working with and designing biometric technology, and they should be discussed by users and by jurisdictions that will legally regulate biometric applications. But I am also fascinated by the possibility of thinking about this crisis in privacy differently.

The desire to be left alone that characterizes many legal and institutional ideals of privacy testifies to the impossibility of being alone (hence the need for laws to govern how we come into contact with one another). If privacy works with the notion that one can and should expect to be left alone, the apparent invasiveness of biometric technology may well make it more difficult to appreciate how biometric thought might amplify privacy into isolation.

A primary operation of biometric thought is to separate the individual from all others, including family, community, ethnic group, and so on. Keeping in mind that this is an uneven practice, such that one can be individualized under biometric thought and subject to prejudice because one's individuality is seen to affirm a racist fantasy, the effectiveness of biometric technology depends upon the imposition of a form of radical privacy that separates individuals one from another. In fact, what happens under biometric thought is the creation of the person as a full-fledged enclave, no longer hoping to be left alone but impossibly alone, first and foremost. Biometric thought produces an ideal in which individuals become private islands, each utterly distinct and absolutely mapped, defined, and verifiable according to biometric cartography. This is a privacy no longer for the individual but a privacy wielded and imposed by the operation of biometric technology.

To put this another way, under biometric thought, individuals become secrets that can be penetrated. Privacy is no longer something one has or has a right to, but becomes something one is. More, this anonymity can become framed as intolerable precisely because it can be overcome, at least in some fashion. As I will examine later in this chapter at length, this process of producing individuals as secrets to be unveiled is an effect of biometric thought that celebrates openness and renders individuals legible, such that one's essence can be meaningfully read upon the surface of the body.

By contrast, consider another island existence that does not produce the individual as a secret to be figured out. J. Hillis Miller writes, almost romantically, of a state he calls being "enisled," in which one "is entirely separate from any community, marooned on a bridgeless, isthmusless island" (129). He is thinking about the singular individual within Jacques Derrida's thought and the tendency of language and culture to overwhelm that singularity with categories that impose upon and attempt to negate the difference of every individual from every other individual. To be identified as a man or an American or a student, for example, is to be transformed from a singular individual into something known and charted and subsequently subjected to norms of how such a person can or should exist. The application of cultural categories almost always erodes the difference of another. For Hillis Miller, the enisled individual is a figure for how one exists outside of the cultural cartography that makes one legible. Such a philosophy of difference aspires to abandon categories of thought that police and produce individuals as knowable entities. This island existence tries to respect all of the ways any one of us may exist beyond the comprehension of another and beyond the social categories that make one legible. Where biometrics dreams of private islands that are carefully charted and identified, Hillis Miller wonders what might be possible if we acknowledge that one can never know another, that another is an unknown island, and that respecting one another begins with this difficult insight. There is no biometric future for such a thought. And given our biometric present, living with such illegibility risks a great deal and may produce forms of privacy scarcely different from privation and unlivable loneliness.

One final thought regarding the crisis that privacy represents in biometric thought. Privacy was once compared to publicity by Kant when he distinguished between the private activities of a civil servant that must reflect one's role and the public duties that are required of that same individual as a civilian. Key to the promise of Enlightenment, as Kant saw it, was the creation of a public sphere in which "the public use of one's reason must be free at all times" ("An Answer" 19). Perhaps privacy is no longer primarily opposed to publicity and the idea of acting in public by addressing others. Perhaps privacy is now also defined by its relation to security and a world defined by the capacity of security to attenuate privacy, a world in which no one may be left alone or permitted to act with unchoreographed freedom.

SECURITY

Social perceptions of biometric technology tend to cluster around a concept of security. When Canada identified its meager goal of accepting 25,000 refugees from Syria before the end of 2015, this was greeted with ready celebration by some and immediately decried by others as a measure that would make the nation vulnerable to terrorist attack. Such a criticism assumes that those fleeing war in Syria are terrorists in disguise, or that all Syrians are terrorists, or that human migration is always a matter of national security. The government responded by invoking biometrics in order to reassure citizens: "Canada will be working closely with the United Nations Refugee Agency (UNHCR) to identify registered Syrian refugees who can be resettled. Canada's focus will be on identifying vulnerable refugees who are a lower security risk. Robust health and security screening will be completed overseas. This will include the collection of biographic information and biometric screening of all refugees, verified against databases" ("Canada"). In this instance, biometrics serves as a vague yet precise-sounding technological means of ensuring safety and security, in much the same way it has ever since September 11, 2001. The attacks of that day marked a transformative moment for the biometrics industry and its claims to secure airports and ports, border crossings, passports, and visa applications. No longer a technology of future dystopias, biometric authentication has been consistently attached to ideas of security. Yet the distance between biometric identification and national security is substantial, especially when we remember that the hijackers who murdered so many on 9/11 entered the United States on legitimate visas.

When the Department of Homeland Security tested airport security in the United States in 2015, they succeeded in getting banned items—including weapons and explosives—past TSA screening checkpoints a staggering sixty-seven out of seventy times (Bradner and Marsh). Such failures remind us that security has become a ritual, and its primary effects may not be to secure at all but instead to create a sense of security, which is also to say to create rituals of insecurity and fear. Perhaps because the problem of security feels so substantial and because political violence almost always arrives without expectation, the promise of identification offered by biometrics is a compelling one because it promises to dispel uncertainty, almost as if it confuses the problem of not knowing with the problem of insecurity. How we recognize what security means and how we organize the world according to that recognition are among the most powerful effects of biometric thought.

Concerns with security tend to affirm vulnerability and authority at the same time. They instill fear as well as a discomforting awe at the extent of efforts to ensure security, whether this is because measures are sophisticatedly subtle or overt and substantial, thereby reflecting in its own scale the extent and scope of an otherwise intangible threat. It may be absolutely right to expect government to engage in tireless security to protect its citizens, but its reach has the effect of suggesting that terrorism is a far greater threat to life and liberty than it has historically proven to be for many nations.

Within biometric thought, security is the imagined consequence of all verification measures. It is the motor for most, if not all, biometric measures. Yet in the name of protecting vulnerable bodies and vulnerable identities, biometric thought produces modes of existence that make individuals more vulnerable and less secure. This statement may seem absurd, because it contradicts what is best understood as the primary accomplishment of biometrics, but security's claims and its practices are often at odds. I don't doubt that biometrics can effectively control access or authenticate identity. But this is not the same as securing something as complex as a person or the nation, matters that cannot be reduced simply to regimes of individual access. Are individuals made more secure within a world in which the majority struggle to survive while others, by accident of birth or luck, have incommensurably better opportunities for themselves? As a domain, security tends not to consider the conditions within which we all live and how they affect us. But perhaps a politics that takes seriously what it means to be vulnerable can and must do exactly this.

Security disenchants the world. It is a mode of justifying unwelcome decisions and procedures. The single outcome of security trumps all else, even if it is little more than a claim to security that offers few measurable results. Security dominates other modes of thought, and alternatives to its logic begin to appear quaint and hopelessly irrational: Aren't security operations justified if they can save lives? How foolhardy and even violent would it be to place others at risk by failing to prioritize security above all else? Now is the time for security, so many say, and once the world is safer, then, maybe, we can afford romantic visions of human movement. There is an entire book to be written on the pleasures of prohibition, including the especially malicious joy of saying "no" on behalf of others. Debating whether or not we should tightly regulate human movement sometimes becomes almost impossible in the context of a logic of security that teaches one to recognize the necessity of controlling how freely individuals move about the globe.

That we treat security as a human concern and look to apply facial-recognition software in order to detect threats shows how limited our ability remains at present to speak adequately about the forces that make us less secure and life on the planet less sustainable and survival more tenuous. The most powerful conditions shaping life on earth are forces that may never have a single human face. Aside from laws designed to track the financing of terrorism, financial circulation in the world is largely unfettered. Here is an example of what this freedom accomplishes. According to CBC Radio One, in 2010, Heritage Oil sought to migrate the company—on paper—from Uganda to Mauritius in order to avoid paying US$400 million in capital gains tax on the sale of property in Uganda. This is more than double Uganda's annual health care budget in a country with 1.5 million citizens who are HIV positive, thousands of whom receive no treatment whatsoever ("Panama Papers"). The Ugandan government pursued Heritage Oil in court and was successful, five years later, in obtaining this tax revenue. Perhaps this egregious attempt is overly dramatic given Uganda's exceptional challenges. Michael Lewis, commenting on Heritage Oil in the wake of the leak of the Panama Papers from the law firm Mossack Fonseca, notes:

> The practice of which they're accused is a practice called treaty shopping. Basically, where you shift a transaction or a company from one place to a tax haven at the stroke of a lawyer's pen. And that's an incredibly common tax avoidance technique. We see it in dozens and dozens of major listed public companies, operating right across the developing world and the developed world. So, it wouldn't in any way be an uncommon strategy. It's unusual to see it expressed as clearly as this in the Panama Papers, that this is definitely for tax avoidance, but it's a very common practice. ("Panama Papers")

The freedom of capital to shop for the best deal is something made possible by wealthy nations. This is not an accident or a loophole no one has noticed. Lewis explains, "We've just been through a round of international tax reforms at the OECD [Organisation for Economic Co-operation and Development], where Canada sits at the table and developing countries are not allowed to sit at the table. And they explicitly decided not to deal with the kinds of treaty shopping tactics that *Heritage* has been accused of" ("Panama Papers"). The OECD refuses, in other words, to police the migrations of capital. The International Monetary Fund (IMF) calculates that when the freedom of capital takes the form of tax avoidance, it costs the global south, which they represent as all non-OECD nations, US$200 billion annually. The cost is more than double that figure for OECD countries (Crivelli *et al*). Lewis notes that this figure is roughly equivalent to the annual global development aid budget. And yet individuals in my state, as in many other parts of the world, immediately and selectively fixate on the security implications of settling and supporting refugees from Syria. There are costs to global migration, and when we are routinely encouraged to notice the human face of global displacement and movement, we may not similarly ask about the human cost of the migration of capital. Security works to curtail the freedom of people but leaves the freedom of capital an unquestioned right (when it even allows discussion of it).

All of this says very little about the effects of investment and the exploitation of particular nations for the benefit of others or how maximizing profits often depends upon sustaining intolerable working conditions by ensuring that populations cannot move freely and seek out better opportunities. The deadly collapse of Rana Plaza, a garment factory in Savar, Bangladesh, in April 2013 stands as an extreme representation of the consequences of this mentality. Employees created clothing for chains such as Children's Place, Benetton, Walmart, and Loblaws. After cracks appeared in the building's foundation, workers were told it was safe to return and anyone who did not do so would lose a month's worth of wages. The next day, 1,137 employees died when the building collapsed. Barely six months earlier, 117 Bangladeshi garment workers were padlocked inside a burning factory because managers refused to believe there was a fire. If biometrics have become a primary technological means and a compelling conceptual force that frames human movement as suspect and dangerous, it is also part of a web of discourse that makes possible the killing force of exploitative labor conditions that relies on immobile populations.

Biometric thought seeks to regulate human movement as if it were a security concern, and as such it relies on the spectacular nature of terrorism to ignore the many more lives abandoned to abusive and intolerable economic conditions. Who benefits from such conditions, and what is their responsibility to those whose suffering makes possible their affluence? As Brown comments, "What the conventional narrative articulates as security issues are often in fact consequences of neoliberal globalization, and contrary to the received wisdom, economic imperatives frequently produce what are characterized as security concerns" (95).

I have suggested that movement may be a powerful response against exploitation and abandonment. What of those who cannot move, perhaps for reasons of infirmity or because their life is conditioned by a bodily existence that is not oriented toward movement or by social attachments that make movement particularly challenging? While millions fled from Syria into neighboring nations and into Europe, how many more could not undertake such a journey? Biometric norms make some lives less visible—in that case, to the extent that they assume that those who need to migrate can and will. A logic of biometric security is all too willing to forget those who will never beat a path to another's border.

BIOETHICS

The Universal Declaration on Bioethics and Human Rights (UDBHR) addresses "ethical issues related to medicine, life sciences and associated technologies as applied to human beings." Accounts of the ethics of biometrics tend to emphasize the principles set out by Article 3, which reads that "1. Human dignity, human rights and fundamental freedoms are to be fully respected; 2. The interests and welfare of the individual should have priority over the sole interest of science or society" (UDBHR). As a mode of thinking about biometrics and making sense of what it does and where one should place emphasis when considering it, the contributions of bioethicists highlight matters such as informed consent, intimidation, function creep, data security, and reliability.

In his chapter on biometrics in the *Handbook of Global Bioethics* (2014), Emilio Mordini considers a representative expression of anxiety regarding "bodily

signatures mechanically extracted from our bodies by impersonal devices . . .
[that] speak of our biology rather than our biography" (514). Where consent
and data security are obvious areas of concern for professional ethics, Mordini
identifies a key feature in the popular perception of biometric technology. He
cites the French National Consultative Ethics Committee for Health and Life
Sciences in order to expand on the implications of a vague public anxiety with
transforming oneself into data:

> Do the various biometric data that we have just considered constitute authentic
> human identification? Or do they contribute on the contrary to instrumentaliz-
> ing the body and in a way dehumanising it by reducing a person to an assortment
> of biometric measurements? Is there not a possibility that this attempt to arrive at
> a biometric simplification, which cannot ever capture an individual's essence, could
> in fact lead to misrepresentation, to seeing nothing but the biometric persona, how-
> ever scientifically determined? They may reduce human beings to an accumula-
> tion of data and cartographic criteria. (qtd. in Mordini 513)

I am less interested in whatever is meant by "authentic human identification" than
in the concerns regarding how biometrics localizes identity in a manner that loses
so much that is key to who one is and how one experiences identity—features
that are never personal and which are locatable precisely beyond the body, in a
social world of others and of cultural norms, beliefs, and attitudes. That is to say,
the scene of debate set out by bioethics—by reading the body, is biometrics a violation of the authentic individual?—already entails fundamental assumptions
regarding the presence of identity in an individual body.

Mordini ultimately dismisses the sorts of concerns noted by the French government and arrives at the following claim, which offers a choice between identification and facelessness:

> People—and sometimes scholars—are victims of the illusory belief that personal
> identification per se threatens basic liberties and infringes the private sphere. To
> be sure, any process of personal identification implies that individuals are recog-
> nized subjects of rights and obligations, and this could be seen as a limitation of
> individual liberty. Yet there would be no rights, no liberty, without personal iden-
> tities. No political, civil, and social right can be enforced on anonymous people.
> One can claim her rights, including the right to refuse to be identified, only if she
> is an identifiable subject and if she has a public identity.

> In ancient Greece, slaves were called "faceless," aprosopon. The word that in Greek
> designates the face, prosopon, is also at the origin of the Latin word persona, per-
> son. The person is thus an individual with a face; this is to say, out of metaphor,
> one becomes a person when she is identifiable. Biometrics could contribute to give

a face to such a multitude of faceless people who live in low-income countries, contributing to turn these anonymous, dispersed, powerless crowds into the new global citizens. (525)

Such a non-choice is ironic coming from one associated with ethics. Should one reasonably be expected to choose between being excluded from social life and human rights, on the one hand, and being subject to biometric scrutiny, on the other? And would we consider this an ethical choice? For Mordini, the person can only be the biometrically recognized person. And what is perhaps most remarkable is that he sees that biometrics will grant faces to those who presently have none. His metaphor of a prejudiced social system that selectively failed to identify the faces of some ("slaves were called 'faceless'") quickly ceases to be a metaphor when he identifies "a multitude of faceless people who now live in low-income countries." What emerges here is a logic in which one recognizes the naturalness of biometric identification but is never permitted to ask after the nature of prosopopoeia, or the art of giving something a face, in the first place. And he does exactly what one expects of biometric thought: he sees equivalent faceless individuals rather than being able to take stock of all the individual differences that make up a population of any nation, let alone all of the differences that are annulled when one speaks of people from "low-income countries." The question that fascinates me is not should one be identified or not, but how does a given process of identification function and what does it do, perhaps in excess of what it claims to do? To give someone a face implies that one does not have one yet. This is a social and cultural operation that reflects a set of priorities and assumptions about faces, identities, social standing, giving, receiving, having, reading, and being legible to others.

HUMANISM

Humanism refers to the idea that human activities and human desires are the source of social life. This perspective tends to be aligned with empiricism, since it grounds its worldview in directly identifiable human activities. Biometric thought might not appear to be a likely outpost for humanist thought, given that it tends to transform individuals into data sets on the basis of impersonal and machine-oriented modes of recognizing individuals. Yet biometrics draws upon a humanist tradition in the way that it focuses on individuals. Its humanizing logic puts faces to names and insists on a human account of violence and trauma, even if it tends toward amassing information on those affected by displacement and dispossession, for example. The image of Syrians and others walking across Eastern Europe in 2015 formed a spectacle of individuals who had made a choice to search out a livable life. Putting a face to hope and loss as well as the separation and reunification of families may make other ways of understanding a humanitarian crisis or social conflict disappear from view. A humanist tradition assumes, for example, that one may be more responsive to faces of suffering than to accounts of structural inequality and failures to protect innocent lives by the international community.

This tradition shapes how identity is apprehended as such and the conditions under which one becomes an object of biometric knowledge. This means that giving a face to something or looking one in the eyes does not discover the truth regarding how things really are but instead reflects and devises the rules according to which a particular truth can emerge.

My analysis of biometrics belongs to what Foucault called a "critical history of thought [that] is neither a history of acquisitions nor a history of concealments of truth; it is the history of 'veridictions,' understood as the forms according to which discourses capable of being declared true or false are articulated concerning a domain of things" ("Foucault" 460). What is true for biometric thought involves selectively seeing some things and ensuring that others fall out of view and thus constraining a field within which truth can emerge, a truth premised on decisions already made about the borders of identity, what migrations can be recognized, and the ways in which we do or do not depend on others and on social structures in order to be ourselves. Such operations are no doubt manifested in human activity, but they may not be reducible to operations of individual desire or will, either. To speak of a social life of biometrics is to recognize that it produces a sense of the world, rather than revealing the truth of it, as if by penetrating beneath a disguise. Rather than seeing who is really there, biometric thought begins first by working to produce a sense of the human, a being who is defined by a face rather than facelessness and that to have a face and to be recognizable is to be identifiable. To lack a face is to be unidentifiable and perhaps not human. The ease with which some can be seen as if faceless, as I will discuss in a moment, should give one pause regarding a tradition of thought trained to see individuals and to ignore the conditions within which they become more or less visible.

This chapter does not offer a complete list of the domains attached to biometric thought. Given the relevance and consequence of biometric thought at this moment in time, it may reach into nearly every area of social life, and thus exhaustively accounting for its attachments may be impossible. Rather than itemize every domain relevant to biometric thought, I have chosen to demonstrate some of its most powerful domains and the social and cultural conventions they involve.

In the pages that follow, I offer two examples in order to consider how biometric thought works by attaching impulses of identification to social concepts, including the ones I have discussed here. One instance comes from a national government, the other from a work of fiction. One reinforces particular norms, and the other tends to expose their operations. What the examples show together is that biometric thought does not issue from a single source and is not limited to discrete acts of identification. It has a social life that organizes and affects how we think about many other ideas, and it frames how we comprehend reality itself.

The first example captures some of the troubling ways biometric thought operates upon individuals and how it seeks to produce them as legible bodies and police how they can present themselves. On December 12, 2011, Canada's immigration minister, Jason Kenney, announced that his government had banned veils such as the niqab during citizenship ceremonies. His comments that "all those taking the oath do so openly" are worth dwelling upon in order to consider how the face has been constructed as a mark of citizenship, identity, and legibility in ways that are complicatedly related to an ethics of obligation and a politics of reading. After announcing the intention to educate would-be citizens on the values held by Canadians and to implement new measures to assess competency in French or English, Kenney turned his attention to the act of taking an oath of citizenship. His comments on this subject form the culmination of his address as well as its most substantial component:

> Effective today, everyone will be required to show their face when swearing the oath. I have received complaints recently from members of Parliament, from citizenship judges and from participants in citizenship ceremonies themselves that it is hard to ensure that individuals whose faces are covered are actually reciting the oath. Requiring that all candidates show their face while reciting the oath enables judges—and everyone present—to share in the ceremony and to ensure that all citizenship candidates are in fact reciting the oath as required by law.
>
> This is not simply a technical or practical measure—far from it. It is a matter of deep principle that goes to the heart of our identity and our values of openness and equality. The citizenship oath is a quintessentially public act. It is a public declaration that you are joining the Canadian family, and it must be taken freely and openly—not with faces hidden.
>
> To segregate one group of Canadians or allow them to hide their faces, to hide their identity from us precisely when they are joining our community is contrary to Canada's commitment to openness and to social cohesion. All I ask of new Canadians is that when you take the oath, you stand before your fellow citizens openly and on an equal footing.
>
> I ask that all new Canadians participate in this ceremony in the same way that you made the solemn commitment to participate actively in our Canadian community. If Canada is to be true to our history and to our highest ideals, we cannot tolerate two classes of citizens. We cannot have two classes of citizenship ceremonies.

Kenney does not specify whose faces are hidden during citizenship ceremonies, though he does very clearly mark them as a unified and willful threat to Canadian values because they would challenge what is considered normal by preventing others from seeing their faces. His comments were immediately understood to be directed at Muslim women who cover their faces in some way. His concern with covered faces might reference the hijab as readily as it does the niqab and is part of frequent identifications that see veiling as a problem in Western discourse because it is a visual interruption in a sea of normatively open faces. I would suggest that such a sense of the veil has little to do with the material object or the significantly different degrees of coverage that different styles of dress provide and much more to do with a fantasy of openness that encodes a very particular biometric understanding of what the face is, what it does, and what it means to insist upon seeing it.

This contemporary imperative toward openness has a history, and this history is complicatedly involved in an ethical rhetoric of face-to-face sociality that has been repurposed to forcibly unveil some people and which re-inscribes the idea that one's face ought to be open and legible. What needs to be said from the start is that the history of such openness is powerfully gendered. Covered women are at issue here, as is a male privilege that insists not just on seeing women but on them making themselves available to sight. A politics of public visibility undertaken in the name of transparency embarks upon a purposeful misapprehension of the relations of power involved in looking, especially at women, in Canada. More, it renders entirely opaque another understanding of public visibility that would discuss instead how individuals are allowed to appear or made to disappear from view. By evoking an empirical account of covered women, this approach produces a particularly powerful fantasy figure: that of the veiled Muslim who appears in Canadian public life as a problem, as someone not sufficiently versed in what it means to be a member of the nation, and as someone unwilling or unable to embody norms of freedom and openness. Such an approach instrumentalizes the lives of real women in the service of an Islamophobia with feminist pretensions, which construes some women as victims of an un-Canadian, backward, singularly patriarchal, and nonprogressive culture.

Noting even just some of what is embedded within Kenney's politics of visibility does little to diminish its effects, given the rhetorical pull of the face as something that is normatively visible within a narrow set of recognitions. I say narrow because I sense that it is not just concealment that is out of bounds but also forms of appearance that make demands or do not accord with convention, such as when women appear angry in public or people of color dare to speak against a culture of white supremacy. Kenney's comments on openness conceal just how untenable many forms of public appearance can be and how strongly they will be denounced before they are ever heard.

Kenney's irresistible language of concealment strategically misrecognizes openness as an expression of "social cohesion" and absolute freedom. Is Canadian society predicated upon a refusal of the separation of individuals that is supposedly marked by the veil, or is it characterized much more by alienating social and economic practices of separation and division that might be spectacularly ignored by refocusing attention on an apparently controversial object? The forced and symbolic unveiling of women and the shaming this law hopes to produce is far from a significant victory for women's rights. Seriously advancing the rights of women is something that Stephen Harper's Conservative government showed little interest in doing, and such pageantry and political theater only trades in illusions of equality.

This law does not engage with the long and varied history of the veil and its relationship to any number of social and personal matters. We might think quickly of questions of gender and power, freedom of choice, visibility and invisibility, social mobility, religious practice, and desire; and we might think more slowly that the veil can also refer to practices of mourning, rituals associated with Catholicism, histories of opposition to imperialism, and even questions of creative self-expression. What may be most substantially and maddeningly clear from Kenney's comments is that this is not meant to solve a problem as much as it is to create and enflame one.

A norm of transparency makes it possible for a veiled Muslim to be cast as a problem, and she is a problem because of how she is both present and absent simultaneously: a threatening presence and an inscrutable strangeness that belongs elsewhere. This discourse has little purchase on the lived experience of what it means to wear, or sometimes wear, the veil or the ways in which women of many religions and nationalities have been working with, behind, and against the veil. It is a discourse that strips away context and isolates individuals from the world and the social codes within which they act, selectively remembering some and not others. As such, it refuses to remember that the veil is much more than an expression of unfreedom.

If there is a comprehensive history of the veil, it is, no doubt, the story of a great many ways of living with and against prescribed social norms, a history that is always lived individually and socially, and a history that ranges across a number of locales in ways that make all the difference:

- The authenticity movement in Iran in the 1970s that "glorified Islam as a social remedy to all social problems" offered the veil as an object of regulation by forces seeking to revolutionize Iran either by removing or mandating coverings for women (Zahedi 257);
- The specter of Algerian resistance and its masquerade of the colonizer's culture—"carrying revolvers, grenades, hundreds of false identity cards or bombs, the unveiled Algerian woman moves like a fish in the Western waters" (Fanon 80);
- Its subsequent reversion in an effort to mask this masquerade "under the protective cover of veils" (Masood 220) was a move that showed dress to be a strategy as much as an article of faith or battleground;

- Its contemporary status as a deeply ambivalent article of clothing means it is worn for reasons that range from "the mundane to spiritual, from the personal to the social" or simply because it provides protection from the sun (Jain 242); and
- It is even something that can undo a person, such that "if you don't look like a caricatured Muslim, you simply cannot be one" whether living in Damascus or Seattle (Masood 225).

This Canadian policy conjures a fantasy of the veiled woman who represents an offensive subservience to men and to Islam and thus becomes a decisive foil to liberal values. As we have seen with the war in Afghanistan, the protection of vulnerable women became, for a time, a primary rhetoric "used to consolidate support for the invasion" among the Canadian public (Khan 168), a rhetoric that was used less and less once it became clear that military intervention created and sustained a situation in which, for example, Malalai Joya, then a parliamentarian and one of Afghanistan's most prominent feminists, was exiled because of threats to her life after she spoke out against corruption in her government. This new policy relies on these past associations and weds them to a biometric logic that suggests the face is a mark of identity and its availability to the eyes of others is a primary expression of one's participation in liberal democracy. Kenney's suspicion regarding those faces that he can only interpret as concealed is far from isolated and recalls the sort of cultural nationalism that led Jacques Chirac, in 2003, to state that "[w]earing the veil, whether it is intended or not, is a kind of aggression" (qtd. in Scott 158–159). Reading France's veil controversies—a collection of cultural anxieties, Islamophobia, and political opportunism—Joan Wallach Scott noted that Chirac's comments were not directed toward cultural codes that encouraged or required women to cover themselves but to the cultural presence of veiled women within French society. If wearing a veil in public could be an expression of aggression, what is the nature of such an act? Scott comments,

> The aggression [Chirac] referred to was two-fold: that of the woman but also of the (Western) man trying to look at her. The aggression of the woman consisted in denying (French) men the pleasure—understood as a natural right (a male prerogative)—to see behind the veil. This was taken to be an assault on male sexuality, a kind of castration. Depriving men of an object of desire undermined the sense of their own masculinity. Sexual identity (in the Western or "open" model) works both ways: men confirm their sexuality not only by being able to look at—to openly desire—women but also by receiving a "look" from women in return. The exchange of desirous looks, the availability of faces for reading, is a crucial aspect of gender dynamics in "open" systems. (159)

Scott highlights the presence of a biometric logic that sees faces to be either visible or inscrutable. And while I am not convinced that faces are quite so open or ever so concealed, I find that her attention to the sexual politics of a Western rhetoric of transparency—a rhetoric that posits the presence of both aggressive concealment and openness—helps to establish the social conventions and codes of meaning that structure an ideal of openness, even if we might both agree that it in no way obviates the complexity of what we apprehend in the face of another.

Biometric thought is uninterested in human complexity, preferring instead to discern simple transparency. It dreams that the face is little more than an instrument of identity precisely by impoverishing the individual of the context, social relations, and attachments that define a concept of the face and who can be said to have one. Not only does biometric thought affirm an uninterrogated moral norm of transparency, but its operations work to withhold so many other ways that would actually enable a serious discussion of what it means to appear and the conditions that regulate how one appears and what that appearance means to others. Biometric thought, in this instance, becomes a way of ensuring that we do not see, or think, about what it means to look at another and what we can and cannot see in her face.

In my next example, I want to consider what it means to think in ways that diverge from biometric thought and ask, Can a work of fiction tease out new ways of thinking about migration? It may seem strange to turn to fiction in a book concerned for the lives affected by very real practices of biometric inspection. The operations of biometric thought are not fictional in any sense, but they share with fiction a commitment to mediating reality by taking up certain aspects of it and not others and by placing emphasis on only some details. Framing what one can see and consider relevant, biometric thought is a way of telling a story about identity. It is a story, moreover, that travels with disciplinary force and the authority to mandate its vision. Fiction is sometimes imbued with that same socially normative ambition, but its ability to insist on its viewpoint is rarely backed by police powers. If the human impacts of biometrics are derived from assumptions and perceptions authorized by biometric thought, then changing how one thinks about such matters may create the social will to alter how biometrics regulates movement on the planet.

Yuri Herrera's *Señales Que Precederán al Fin del Mundo* (2009), translated into English in 2015 as *Signs Preceding the End of the World*, is a novel that departs from a number of the assumptions that guide biometric thought and its mode of mediating and imagining reality. The slender yet expansive novel plays on the notion of the world as a spatial, conceptual, geographic, and lived reality. The end of the world, in this case, marks the beginning of another. The novel comprehends "the world" as the legible product of norms and ideals that regulate how one sees reality and which is subject to change depending on one's culture and perspective. Such world-making properties are at the core of biometric thought and its capacity to condition how we think and live within a conceptual environment that can be hostile or supportive to different forms of living and the desires that make us who we are, even if its conceptions and authority are never absolute.

The topic of how biometric measures of identification police thought is central to Herrera's story about an indigenous Mexican teen, Makina, who travels clandestinely, though not unnoticed, across the desert into the United States in search of her brother. It must be said that the novel has almost nothing to say directly about biometric technology. Instead, it emerges as the unconscious of the book. Makina continually marvels at how those who have been secured to help her cross the border and aid her on her journey readily recognize her. Hers is a world of human identifications rather than biometric authentication. Strangers unexpectedly recognize one another as familiar; meanwhile, Makina almost does not recognize her own brother, so transformed is he by his new surroundings. In Makina's world, identity can be falsely assumed, bought and sold, violently ignored, and lost in ways that make the assumptions of biometric thought seem strange and fantastic. What fascinates me, then, is how powerfully this account of individuals traversing borders and unmaking themselves—and the world— departs from the impulses associated with biometric identification; it captures the richness of ways of being in the world that have been forgotten by biometric thought.

Makina's brother was drawn north across the border by the suggestion that the family might own some property in the United States. When he sends increasingly evasive notes back home, their mother asks Makina to go and find him and bring him home. What follows is a story of identity's alteration rather than its permanence. It is also a story about recognition, or the social process of making sense of a world that is guided by norms and conventions that structure how we think about and perceive the world. Before she locates her brother, Makina finds the property he searched for. If this land was, for her brother, the external cause for an internal search for identity and his future, what Makina finds there is a territory that looks like nothing but loss. Not only was the land a "sheer emptiness," but the excavators "were still at work" removing all trace of the particularity of this land (69). It was "as if they needed urgently to empty the earth; but the breadth of that abyss and the clean cut of its walls didn't correspond to the modest exertion of the machines. Whatever once was there had been pulled out by the roots, expelled from this world; it no longer existed" (70). What is this excessive operation that annihilates the life that existed there, the excavation that eradicates every trace and memory of what once was? There is no typically biometric connection between surface and depth here. This land is no longer recognizable as itself once its outward manifestation no longer exists. But perhaps this is what biometric thought does when it excavates the self and makes unthinkable all those other ways of knowing oneself and existing in the world, including attachments to the land and to others, pulling out the roots and leaving one exposed but barely recognizable.

Makina is jolted out of these reflections by someone nearby: "I don't know what they told you, declared the irritated anglo, I don't know what you think you lost but you ain't going to find it here, there was nothing here to begin with" (70).

Herrera offers the reader at least two ways of thinking about the ground beneath these characters' feet. There are those who seek to eliminate the ground and scrape it bare. And there are those who go looking for the land. Some seek to forget the roots that trace a history beyond the present. And others seek to remember what was there and what it meant. Makina discovers razed ground but still sees the signs left by those trying to destroy even the traces of the destruction wrought here, erasing all that might have made this place recognizable to her family. The loss of this land is not in doubt. But what land is becomes a question haunting Makina: how does one live when some would try to insist that the earth is nothing and never was anything and that we exist without it?

When Makina finds her brother, he tells her the story of how he has come to be in the U.S. Army under an assumed identity. Able to survive thanks to the hospitality of others, he was approached by a woman who asked him to take her son's place in the military. "This is who you are going to help, said the woman. But I want you to meet the whole family you'll be saving" (87). Individuals are never alone in Herrera's book, and saving one means saving many more from so much pain and loss as well. In exchange for assuming her son's place in the military, "the family would pay him a sum of money. A large sum, they specified. Plus he could keep the kid's papers, his name, and his numbers. If he didn't make it back, they'd send the money to his family" (88). After a stint of three months' service abroad in a war that Makina's brother will not speak of, he returned to the family. After congratulating him on his service, they "thanked him on behalf of his country," and the father "began to stammer something about how hard it was to get the money together and how complicated it would be for Makina's brother to use his son's identity and about the possibility of him working for them instead, and that way, if he wanted, he could stay in the country legally. But the mother didn't let him finish. Said No. Said We're going to keep our promise . . . we'll go someplace else, the mother replied. We'll change our name, reinvent ourselves" (91). The narrative of Makina's brother recasts so much of what we think we know from biometric thought. As a newcomer, Makina's unnamed brother is not "an illegal" nor is he seeking to falsify his identity. Instead, he helps an entire family by assuming their son's role in the military. Makina's brother becomes a model citizen, one who serves his country and helps others survive. The family assumes the role we might have thought only possible for Makina's brother: they must forge a new identity in a new place and continue to test their resourcefulness. Their paths are not equal, and the father's reaction reminds us how vulnerable Makina's brother is to abuse in this situation, but each share a similar relationship to identity, movement, and survival. This conclusion is not one biometric thought and its border patrols conspire to produce. Gone is any moral logic of deception and transparency, replaced by the possibility that life depends on legal identities that can shift from one person to another and that this is a matter of obligations and duties to one another. Identities may not be ours and might, variously, threaten families as readily as they make another's life more livable; be occupied strategically; or be abandoned and lost. They are not attached to bodies but are themselves capable of remarkable acts of migration.

Even official representations of selfhood can move about and have a certain indifferent autonomy such that they can represent different things to different people and may be easily left behind or preciously guarded. Though he wishes to remain in the U.S. Army, Makina's brother had "no clue what to do" as a soldier and no reason to remain one, except to act on an almost traumatic faith that "there must be something they fight so hard for. So I'm staying in the army while I figure out what it is" (93). Sometimes the very foreignness of a thing can be the basis for attachment, and this might be the closest the book comes to embracing biometric logic, which likewise depends upon strange attachments between what may be so alien to oneself, such as the pattern of an iris, and a cultural understanding of oneself as a unique human being.

Makina leaves her brother behind to continue his life in the military, and shortly thereafter she finds herself detained in a vacant lot along with others suspected and therefore identified to be undocumented migrants. A self-identified "patriotic [police] officer" lectures them on civility and then sets out to humiliate a man who possesses poetry but no identity papers (98). He orders the poet to write a confession of his crimes, to which the poet cannot respond. Makina grabs the pen and paper and composes the following for the increasingly disconcerted cop:

> We are to blame for this destruction, we who don't speak your tongue and don't know how to keep quiet either. We who didn't come by boat, who dirty up your doorsteps with our dust, who break your barbed wire. We who came to take your jobs, who dream of wiping your shit, who long to work all hours. We who fill your shiny clean streets with the smell of food, who brought you violence you'd never known, who deliver your dope, who deserve to be chained by neck and feet. We who are happy to die for you, what else could we do? We, the ones who are waiting for who knows what. We, the dark, the short, the greasy, the shifty, the fat, the anemic. We the barbarians. (99–100)

Reading this note aloud, the cop drops to a whisper as he finishes it and is so shaken by what he has read that the degrading spectacle of identification and possible detention disappears from his consciousness. His appetite for carrying out such senseless violence dissolves. The police had demanded nothing less than a narrative account of the poet's criminal existence in this territory. Yet what Makina provides is not an account of herself but a powerfully social account of existence and what it means to live insecurely and in the wake of all other unchosen norms that govern how she and so many other undocumented migrants are forced to appear in American consciousness. She refuses the biometric imperative of isolated selfhood and so-called personal responsibility and instead insists that whoever she might be seen to be in the eyes of the cop, such a way of seeing does not reflect her own understanding of herself and barely accounts for the actual relations of power that make the lives of many in this parking lot less livable despite the contributions they make to a world unwilling to acknowledge them. Makina writes the world anew, reimagining the norms and ideals that govern a worldview that seemed so easy, obvious, and untroubled by the possibility that its regimes of veridiction might not be the only truth imaginable.

This scene sounds impossibly hopeful for the power of language to affect how individuals live and act in the world. Biometric thought might rely on narrative modes of framing and representing the world, but it also entails tactics and policing practices that bring force to bear upon its manner of dictating what will and will not be recognized to count as the ground for identity. The difference is not just force but scale. A single sheet of prose cannot muster the same influential gusts as a network of ideas reiterated and re-inscribed millions of times over and buttressed by institutional practices, strategies, requirements, and penalties. But maybe that is the point. Writing can have an effect at a human scale, just as it does here. How we narrate our existence, and whether we allow it to be narrated for us or forcefully insist on telling our own stories in our own ways, may help to determine not only who we are but also the nature of the world that surrounds us. And telling new stories that reinvent the world as Makina does here or continuing to tell those stories a society is unwilling or unable to hear may make lives better even if they come short of entirely changing the world and all of the social relations that make it possible.

In the novel's final chapter, Makina receives forged identity papers that will enable her to stay in the United States, at least for now. It takes place within the sort of impossible spaces that so often populate magic realist narratives in Latin American fiction and which testify to the ways any one of us can confront worlds that seem fantastic yet may be ordinary for those who inhabit them. Makina "entered a little maze of streets that looked like they belonged to some other city and stopped before a low narrow door behind which nothing could be seen" (104–105). Crouching to fit through this tiny door and descending four times around a spiral staircase, Makina arrives at an underworld capable of producing forged documents. It is a bizarre scene that likewise asks readers to consider some of the ways that their own all-too-familiar realities may actually be fantastic, too. This uncanny "place was like a sleepwalker's bedroom: specific yet inexact, somehow unreal and yet vivid" (105). The perplexing reality that Makina finds herself encountering at the close of the novel asks readers to consider the possibility that if something appears impossibly strange, it reflects how we see the world at least as much as it reflects what we see.

This unreal place in which Makina acquires documents to go with her new identity features a single word above its entrance: "Verse" (105). One of the most striking artistic features of the novel is Herrera's use of the term *verse* as a noun turned verb. As the novel's translator, Lisa Dillman, notes "to verse" is her way of translating the neologism *jarchar*, a term that means approximately "to leave" and which can be traced to much older Arabic and Hebrew poetry in what is now Spain (112). What is key to understand about "versing," I think, is that it places emphasis on leaving and the way in which departure is necessitated, structured, and understood. Makina "verses," just as her brother did. She never arrives. She does not enter. She does not dream of America. She leaves.

This may be the most profound contribution the novel makes to thinking in ways that challenge biometric assumptions. Under biometric thought, individuals arrive. They are screened for entry or they have evaded screening. It is a logic that leads to a number of subtle and overt conclusions. It creates an idea that foreigners come here and that "we" better watch those who arrive. It suggests that here is necessarily not composed of people from elsewhere. This is a mode of historical forgetting that ignores past migrations, histories of colonialism, and genocide in the new world. It is likewise a means of erasing the presence of immigrant communities and their contributions to civic and social life in the present by insisting on marking some as out of place. More, it is a mode of thought that always insists that here, wherever it is, is desirable in the eyes of others. Here is preferable to there; here is just plain better. I had this assumption so forcefully revealed to me recently, hearing a Syrian refugee speak about arriving in Canada and how largely indifferent he was about his new land, thankful though he was for the hospitality shown to his family. He did not desire to come here, however; what he needed desperately to do was "verse," to get away from the war zone that his home had become. The implied superiority of here can also stall questions about the inequality, discrimination, injustice, and other dispiriting features that can await newcomers and erode the lives of citizens alike.

Versing is a mode of departure, not arrival. As an expression of leaving, it is not necessarily melancholic. Nor is it attached romantically to a vision of what it is to come. Characters verse because it is necessary, right, or desired in the moment. Versing names movement without a sense of where one will arrive and what the future may bring. Versing does not know what comes next. It may not be a longing for a better life or a recognition that one culture is barbaric and another civilized. It is an expedient option and perhaps the only viable decision to make. But it is not an arrival at a destination or a culmination of a personal narrative. As one individual responds to Makina's question of whether or not he likes living here in the United States, "Tsk, me, I'm just passing through"—even though he has been here "going on fifty years." This is the start of what a worldview based on versing would look like. One has never arrived. One has never fully left.

Biometrics is a logic of arrival and destination, a mechanism for managing access that can scarcely comprehend the way lives are lived back and forth across borders, among individuals who do not seek to get to the other side but instead live in transit, versing back and forth. How strange and different *versing* sounds from *migration*. How different a conversation about human mobility becomes when we allow for the possibility that desire and destination may not be the powerful categories we think they are.

Versing is also a way of speaking, if we recall the work of a poet or a songwriter and the way words leave oneself in the form of verse. Communication too is a mode of departure. It is a mode of leaving oneself, transforming for better and worse what one can into words. It is an encounter with others, but it is also an encounter with social constraints that govern what one can say and imagine or what another can hear and comprehend. Communication is always being made possible and impossible. Versing captures a sense that we can be made and undone by the words that we speak and that others hurl at us, as Makina knows when she pens her reply to the cop, spitting back the violent truth that has been thrown at her. Words leave her but they can also compel us to take leave of ourselves, see ourselves as others see us, seeing ourselves in ways that are not just unrecognizable but also violently undoing precisely because we can be forced to recognize ourselves in terms defined by the desires, hatred, and indifference of others.

Signs Preceding the End of the World does not provide a map of practical steps that might replace biometric procedures. What it shows, instead, is that biometric thought naturalizes a particular set of ideas and modes of thinking about matters such as migration, security, privacy, community, and identity. What cannot be understated is the extent to which a given way of seeing the world is not natural or inevitable but a concerted effort to understand it in this way. There are other ways of thinking about these matters, as Herrera's novel captures so powerfully. *Signs Preceding the End of the World* refuses to accept the disenchantment that biometrics insists upon when it sees threats at every turn and insecurity as the inevitable cost if one refuses to reduce our shared existence to forms of identification and the regulation of human movement.

We lose so much of the world in biometric thought, and if Herrera helps to see some of the dignity we have lost, he also shows how maddening it can be to confront a way of thinking that aspires to lose the trace of that loss as well.

What both of these examples differently document is the extent to which biometric thought has become a primary way of apprehending the world, especially in North America and Europe. Some of our most pressing social conflicts and humanitarian demands are closely connected to biometric thought and the ways in which it makes areas of social life and human existence visible and invisible. Biometric thought makes it possible to see some things and not others. It frames the world, and these frames are expressions of relations of power, supported and supporting practical regimes of access control and identity verification.

4 · ON METHOD

In the preceding chapters, I have suggested that biometrics should be understood not just as discrete technologies of identity verification but also as a set of ideas and ideals that emanate from a range of social, governmental, and cultural sources. As a mode of thought, the social life of biometrics is especially attached to particular domains that it intensifies and which intensify it. In this chapter, I will consider how biometric thought functions to produce and naturalize social recognitions regarding individual existence and identity. I frame this as a discussion of methodology because my attention to the social recognitions made more possible, viable, and, indeed, more likely by biometric thought are a key methodological distinction that sets my argument apart from the work of other scholarly accounts of biometrics. I elaborate on these distinctions at the end of the chapter. First, I wish to consider what it means to take as my method a mode of recognition that might appear to be fundamentally similar to the functioning of biometric technology. What I hope becomes clear in my discussion of biometric recognition are how it brings to light matters that are rarely noticed by the routine operation of biometric practices of identification.

I have considered a range of instances of biometric thought that establish a social preoccupation with identification in a variety of pockets of culture, including fictional and nonfictional narratives, scenes of official identification, visual and performance art, human rights reports, and government initiatives. These instances of identification involve powerful attachments to ideas of human mobility, borders, acts of looking, forms of knowing, fantasy projections, assumptions about the physical self, and relations of power. My examples have often been empirical and concrete in an effort to highlight the great diversity of operations that I wish to suggest can and should be understood to work in concert, over the past two hundred years at least, to regulate how we have come to understand biometric practices and promises of identification. My approach may appear similar to the operations of biometric technology, especially when I address human histories. I would seem to confront identifiable beings and verify a certain truth on the basis of seeing them and recording their presence within a certain context. One might even go further and wonder if this isn't also a form of curious access control on my part: to appear in these pages, one needs to have been prescreened and confirmed to have something to say about biometrics. Even my examples drawn from fiction tend to see representations of individuals as if they were human, prompting a perhaps strange coincidence between what is alive and what is not, as if to reproduce in altered form those morbid realizations that even the nonliving can sometimes be used to gain access to biometrically protected objects.

I have addressed discrete empirical accounts of what it means to live in a biometric present (and this is a present that has been developing for some time and has existed in different ways for different individuals and places and eras) in part to provide a counterpoint to biometric ambition and the often limited narrative of biometrics that would see it as simply a means of technical identity verification or access control. But my approach is not merely one that sees humans and identifies their unique experiences with biometrics. Where biometrics seeks to record presence and match it to a documented record, my method seeks to do more than once again make people countable, identifiable, and verifiable. It seeks to understand how it comes to pass that one can see a person in the ways we often do, in the first place.

There is clear value in seeing rather than not seeing individuals and recognizing their existence, and sometimes their deaths. This became immediately clear in the aftermath of the September 2015 publication of Nilüfer Demir's heart-wrenching photograph of three-year-old Alan Kurdi washed ashore along the Bodrum Peninsula in Turkey. Alan and his brother, Galip, along with their mother, Rehana, died when the boat they were on capsized in the Aegean Sea. They were Kurdish migrants attempting to reach Europe and perhaps join family in Canada. How does biometric thought structure what this image came to mean? I want to consider how this powerful image can reveal the operations of biometric thought and then consider how this analysis can illustrate my method and approach to biometric thought.

As a photograph of a migrant and a refugee, Demir's image made visible the lives of so many other migrants and refugees who have left unendurable circumstances. It also marked the ways in which reality becomes recognizable and felt, such that this image moved many parts of the world, at least temporarily, to open its borders. This was a photograph that worked as if by a biometric logic, then, identifying not just Alan Kurdi, but also the migrant crisis in the minds of those who had not yet seen or recognized the desperation that drove so many migrants to risk death as they traveled across the Middle East, North Africa, and Europe in the past decade and especially since the start of the war in Syria. It gave a face to events that many seemed not to see or identify as a humanitarian crisis.

Prompting a humanitarian response, the image identifies an ongoing catastrophe in the form of a single human being's death. This image helped to shift Kurdi's death out of what Berlant calls the "temporalities of the endemic" (97), in which the loss of life is routine and ordinary among migrants. The Dutch

NGO (nongovernmental organization) United for Intercultural Activism has compiled a list with over thirty-five thousand names that documents and remembers the often unidentified individuals who have lost their lives migrating to Europe between 1993–2008. This comprehensive document does not appear to possess the same public power to make these lost lives recognizable, however, at least when compared to Demir's image and its similar refusal to permit the deaths of migrants in the Mediterranean to become unremarkable.

The suggestion that the image of Kurdi may have prompted a humanitarian response by framing the migrant crisis as a matter of suffering individuals is borne out by a systematic analysis of Canadian media coverage conducted by Rebecca Wallace, who notes "that depictions of refugees shifted with the emergence of the Kurdi photo" (226). "The findings of this analysis remind us that news media, alongside politicians and the general public, have the capacity to re-humanize the coverage of refugees by dismantling stereotypes that are driven by threat and fear" (226). A representation has the power to humanize and personify reality. Such a discovery is fundamentally biometric in the sense that it relies on a representational index to document the reality of one's existence.

As Ayobami Ojebode notes, this image belongs to a genealogy of images that have interrupted the regular unfolding of terrible events by giving a face to a particular reality, recalling as it does images from Nazi Germany, Vietnam, Tiananmen Square, among others:

> Major international news media turned news attention not just on the story but also on the power of the image of a lifeless boy to alter the course of refugee and migrant discourse in Europe and America. For instance, in a blog headline, the BBC asks: 'Has one picture shifted our view of refugees?' (BBC 2015). The Independent seems to answer: 'If these extraordinarily powerful images of a dead Syrian child washed up on a beach don't change Europe's attitude to refugees, what will?' (*The Independent* 2015: n.p.). CNN's Paula Newton explains 'How Aylan Kurdi changed Canada' (Newton 2015). But do images really change our world? (116)

It strikes me that these interrogations regarding the power of a public image pose fundamentally biometric inquiries that test the difference that context makes for such moments of visual inspection: How does one encounter a mediated representation of a human, and what does one see when one does? The face can be examined as part of a ritual of official inspection, viewed, recorded, matched in order to control access. A face can go viral and alert people around the globe of a humanitarian emergency. And in some contexts, Kurdi's image might even be received favorably, as painful as such hatred is to contemplate. These operations of biometric assessment and recognition are distinct and serve very different ends as well as reaching vastly different audiences. The official and the public circulation of faces can involve overlap too, in the sense that these very different scenes are sometimes defined by similar assumptions regarding the importance of visual inspection and the assumption that one receives significant information by scrutinizing individuals as if they were objects to be seen, assessed, and decided upon. The meanings that each draw out of a face differ, even if they concur that faces are legible and meaningful. Context matters. The image of Kurdi was not simply appraised by the public as a migrant but as one who deserved hospitable reception rather that impersonal identification. Ojebode sees that context matters too when he asks a provocative question about another audience of this image. He recalls a question like "Do images really change our world?" when he wonders about other migrants who encounter images of Kurdi: "Why do [such] images fail to change the world(views) of these desperate migrants?" (116). The meanings generated by biometric recognitions depend on context. They are not absolute or consistent despite sometimes being static and objective records.

If this image moved the world to see here what Wallace calls "a humanitarian catastrophe requiring support," it was also used online to "express considerable backlash against refugees. In their analysis of Twitter and the hashtag #refugeesNOTwelcome, Rettberg and Gajjala (2015) find that antirefugee campaigns on social media disproportionately focus on male refugees, erasing the experiences of women and children and conveying male Syrian refugees as rapists and terrorists. Depictions of 'predatory sexuality' and 'undisciplined aggression' (180) continue to fuel threats regarding security and reinforce the dehumanization of refugees in news coverage" (Wallace 212). In a compelling polemic, legal scholar Nadine El-Enany wonders if the lines between a racist and a humanitarian response to the image of Kurdi are necessarily so different: "What was it about the photo of Aylan Kurdi that so galvanised Europe's public over the question of refugees?" She notes that he "was by no means the first child to drown en route to Europe," and yet this death and this image came, "all of a sudden, to humanise the body of a refugee" (13). Kurdi came to be recognized by white Europeans, she suggests, as one of "their own (Constanti 2015) sons (Blackwell's Mark Blog spot 2015) and nephews (Mumsnet.com 2015) in the photo (New York Times Magazine 2015), aptly illustrated by the #CouldBeMyChild (Grierson et al. 2015) hashtag, which was trending on Twitter following the discovery of Kurdi's body" (13). She notes this image might come to be filtered through a racist logic that sees "the body of a light-skinned child that enabled the temporary, fleeting awakening among white Europeans to a refugee movement that long-preceded the media spotlight on that photo" (13). The possibility that a humanizing recognition is conditional leaves her to wonder,

What of the refugees who do not evoke in the mind of the white European an image of their own offspring? The images of black African bodies (Sim 2015) washed up on the shores of Europe's Mediterranean beaches last spring did not prompt an equivalent outpouring of compassion and charitable action. What of the bearded male refugee? What of the woman in the hijab or burka? What of their dark-skinned children? These coded images of Muslims inhibit their humanisation. The Islamophobia that thrives in European societies today means that rather than compassion, they elicit feelings of apprehension and fear. (14)

Such speculations regarding the humanization of Alan Kurdi highlight the durability of well-worn paths that sustain white privilege and assumptions regarding the human face of those who have always been recognized to be human in the global north and who have only selectively granted that recognition to others over time. What I find especially striking in this argument regarding the variable and selective way in which humanization functions is that it depends on seeing faces and assessing what one sees there. But this is more than just a matter of recording what is visibly present. El-Enany invites one to also see what conditions visibility. Indeed, what El-Enany shows is that it may be tremendously important to see what is absent from the photograph: audiences and their assumptions powerfully shape what one can see here, yet these are nowhere visible in the image; practices of humanization produce similarity as an index of respect and care, but this is likewise not something one can see on this beach. Is this a fundamentally biometric recognition that aims to see the human at the center of this image, or is this something else entirely?

My argument has been consistently premised on the idea that what biometrics says it does—verify identity by matching records—may not be all that it does. As a mode of controlling access, biometrics involves a recognition: Is this person presenting biometric data a match to a previously documented record of biometric data? This act of recognition is fundamental to the success of a biometric strategy. And problems can arise here. They are the sorts of problems that Magnet describes when she cites a 39 percent failure rate for facial recognition at Logan Airport in Boston (26). They are also the sorts of problems that led Aaron Peskin to introduce legislation in the city of San Francisco that would ban police from using facial recognition technology because it simply does not work as effectively as many believe: "We know that facial recognition technology, which has the biases of the people who developed it, disproportionately misidentifies people of color and women. This is a fact" (Gaiser). One wonders about the failure and success rates that are never mentioned in increasingly ordinary encounters with facial recognition, such as when ISM Connect was contracted in 2018 to provide security services at Taylor Swift concerts. I have in mind a different set of concerns and outcomes regarding practices of recognition that are different from what these engagements with biometric technology address.

To what extent is recognition a process that extends far beyond such discrete activities, and how are its operations coded? Under the logic of operations of biometric thought, Demir's photograph of Kurdi can be identified, perhaps even must be identified. The image is not permitted in any of the instances I have considered to be beyond meaning, as if it could be allowed to stand as a silent expression of an existence lost or memorialize the sheer difference of another whom I did not know. It always arrives first as an identified image of Kurdi, and this begets subsequent identifications about what a visual record of his death can mean. The image comes to be matched to a reality. Giving a human face to reality and making it meaningful involves recording some things and leaving others out. This is the truth of every identification, whether that involves a fingerprint or a photograph. The biometric thought that structures so many responses to Demir's image of Kurdi sees the person first and then sees Kurdi as the face of a larger set of social concerns. To see Kurdi and to match him to a particular reality involves making decisions regarding what this image can be said to record, what it might be said to capture within its frame. And by insisting on this point, I am departing from the logic of biometric technology and the untroubled obviousness it insists upon when it sees a face or a gait or a retina as an index of a person. There is an operation of recognition at work here, I argue, a mode of selecting and seeing that is worth thinking about carefully. This idea that biometric thought recognizes some things and not others when it sees, records, and verifies identity is the basis for my assumption that biometric thought does things in the world by shaping how we perceive ourselves and the world of others rather than simply recording and counting things that punctually exist.

One of biometric thought's most powerful operations was immediately evident when viewers recognized Kurdi as the face of a larger phenomenon of mass migration. Kurdi became recognizable as a migrant, first and foremost, apart from and a part of a larger cultural presence that sees migrants, refugees, and undocumented persons as undifferentiated and anonymous masses. The humanization of Kurdi is a strange mode of identification that paradoxically conjures anonymous crowds in order to dispel them. One must insist on this point, I think: the singularity of every individual was never in doubt, but it is not always recognized, either. Faceless crowds are often implied by biometric promises to produce identity out of anonymity. Existence is not enough, biometric thought insists: one must be seen and identified, and this means becoming legible as a discrete individual emerging out of a condition of faceless existence. The figure of the crowd matters because it can appear to be the basis for so much of the fear associated with migration that comes to be dissipated by the recognition of Kurdi's singular humanity. Biometric thought produces the very problem it promises to alleviate, in a sense. Indeed, part of the power of this image comes from its capacity to recognize the existence of a single person separated forever from others.

If this was the face of a crisis in migration, how are responses to that crisis framed, such that what is required in this time of crisis is the establishment of regimes of order and registration and tracking? Biometric thought tends to construct human mobility largely as an object of uncertainty, as too much, too chaotic, and even a threat to a host culture and way of life. It is an event defined by the arrival of unknown individuals who must be made known, monitored, and determined to be safe. Such perceptions make life more difficult for those who arrive by authorizing and legitimizing discrimination, and they blunt any appreciation for the opportunities that migration brings, as the street artist Banksy highlighted at the Calais refugee camp in France with an image of a migrant carrying an Apple computer. This image starts a conversation regarding the legacy of past Syrian migrants such as Steven Jobs and their contributions to the wider world. Even more powerfully, this image makes a claim on behalf of those present in the camps that their future may yet be characterized by hope and opportunity, not just inhospitality. There is more than one way to see this reality, Banksy insists. And as important as that is, it would be a mistake to not notice that the force of graffiti to record that reality is not nearly on par with the power and influence backing biometric technologies, which shape this reality so differently.

Biometric thought produces a partial understanding of the world, as if looking through a window that constrains sight as much as it enables it. As Roland Barthes once noted, windows enable us to see a particular vista, but they also prevent us from seeing what lies beyond the scope of that window (122). Biometric thought functions in a similar fashion. It frames the world and sees only certain details that compose reality and human experiences. Biometric thought takes firm hold of reality, displaying and filtering it by making some matters powerfully visible while withholding others from view entirely. One can look through this particular window on the world to see not just what is pictured there but also the transparent glass that conditions and makes sight possible. His metaphor is not perfect for this instance, it is worth adding, because there is more at work than just framing and seeing the glass that makes our perspective possible under biometric thought. Arguably, biometric thought develops visual codes that may not be well represented by the figure of a window precisely because they involve excising parts of reality. Perhaps its modes of recognition share more with Photoshop techniques: if biometric thought looks as if through a window and frames reality by doing so, it also has the capacity to select and remove the trees and buildings and people in the background until all we see is a person in the foreground, such that nothing else exists.

Barthes's metaphor is imperfect also because it does not quite explain how, given a particular way of framing reality, we recognize what we see there in the way we do. Why should an official see a head-and-shoulders photograph in my ID and not be suddenly amazed that the absent appendages and lower body recorded by that ID have regrown? I am being hyperbolic, but it is a point that one of the leading theorists of disability studies, Lennard J. Davis, has asked about classical sculptures when he notes that we never see the Venus de Milo, for example, as a representation of an individual whose arms have been amputated (128). One has learned, and is always being taught, to make specific recognitions that may bear little correlation to what is actually brought into view. Recognitions rely on cultural scripts and social knowledge that make it possible, or sometimes mandatory, to recognize reality, and indeed oneself, in particular terms. This point bears emphasis because Barthes's model may give us the impression that the world is presented to us as passive spectators when in fact the most powerful social recognitions include those we make ourselves and those made about us by others, all of which are guided by social knowledge, convention, and norms that shape how any of us perceive the world before us. To return to the figure of the window, if what we can see is constrained by norms of recognition that govern how one sees and processes what can be said to appear there, what we can see is likewise constrained by normative ideas of what looking involves. Is reality, for example, evident to the naked eye as opposed to something that ought to be revealed by a microscope and its very different modes of perceiving? There are almost always multiple ways of apprehending reality and seeing what is there, and social norms guide our perceptions by encouraging some ways of perceiving, discouraging others, and perhaps even making some unthinkable.

Recognition belongs to a history of thought that prioritizes the social basis of existence and insists that the operation of social norms is central to how we exist. As Judith Butler and Athena Athanasiou note, "recognition is not the same as self-definition or self-determination. It designates the situation in which one is fundamentally dependent on terms one never chose in order to emerge as an intelligible being" (*Dispossession* 79). Recognition can be something granted or withheld, demanded or assumed. It points toward the ways in which each of us is dependent on norms that govern how and in what ways one will be recognized. It suggests that my sense of self is conditioned by terms I do not author or control. To be recognized is not the same as existing in the sense that one might exist but not be recognized to exist or not be recognized to matter to the same extent that another matters. Recognition names, then, an attempt to understand how existence is mediated and conditioned by social norms that can produce different possible recognitions of what is actually in existence. It is a mode of seeing that can routinely forget to see some of what is present, and it possesses a disciplinary force in which the norms it establishes or invigorates police how individuals live.

The development of these ideas is often associated with the nineteenth-century German philosopher Georg Wilhelm Friedrich Hegel. For Hegel, recognition names a process that is reciprocal in nature and a foundation of human consciousness. Consciousness, he contends, depends upon the recognition of others as conscious individuals capable of recognizing my consciousness. One is never, in other words, standing in front of the window, alone, making recognitions of the world. To the extent that recognition is possible, it means I have already recognized the existence of others as the basis of the possibility that I can have my existence confirmed and can recognize things about the world. The Italian philosopher Giorgio Agamben contends that "to be recognized by others is inseparable from being human" and that "it is only through recognition by others that man can constitute himself as a person" (46)—if indeed this is best described as an act of self-constitution at all, dependent as it is on the actions and apprehensions of others. Butler's ongoing development of the concept of recognition emphasizes Hegel's notion that it entails what is other than the self and that its force is at least partially reciprocal and thus never wholly determined by a power imbalance in which one side has all and the other none. The reciprocal nature of recognition means that there is never one who recognizes and one who is recognized. Power is never thought of in absolute terms in this context. Even in circumstances in which recognition is withheld or refused, not being recognized does not likewise mute the demand for recognition. For example, the "connection with nonhuman life is indispensable to what we call human life. In Hegelian terms: if the human cannot be the human without the inhuman, then the inhuman is not only essential to the human, but is installed as the essence of the human. This is one reason that racists are so hopelessly dependent on their own hatred of those whose humanity they are finally powerless to deny" (*Notes* 42). Butler's comments highlight the ways in which a self always depends on what is not the self in order to be itself. She also helpfully notes that recognition is unstable and may not be neatly or predictably governed by desire. What kind of humanity is this that is premised on producing inhumanity? Does the recognition of another as inhuman reveal a fundamental vulnerability and need for another at the core of oneself?

My own deployment of the term *recognition* similarly assumes that recognition is profoundly consequential and forms the basis of the social existence of individuals. Recognition is not something accomplished just once, perhaps received as a child, and locked into place. It continues throughout one's existence and continues to shape how we think and act in the world. I follow Butler by noting that every social recognition, every way of trying to order the world and produce, record, and enforce knowledge about it, is partial and always incomplete. If recognitions could enforce their perspective in the world, they would not be so ubiquitous or so insistent. Because the social recognitions of biometric thought remain always unfinished, they are also open to critique and revision, though perhaps not without altering the relations of power that make particular recognitions possible and compelling. Biometric thought insists upon visibility as a moral good, casts migrants as swarms of people overwhelming borders and nations, and identifies so-called illegals as public dangers rather than human beings. If all of these matters were true and reflections of a fully present reality, it would not be necessary for a society to insist upon them so vociferously and continue to make such recognitions again and again and again. There is something revealing about a norm that is so anxious to prove itself to be true.

The recognitions that biometric thought offers are not absolute, then, nor are they beyond critique or revision. Rather than mandating ways of seeing the world, biometric thought makes some social recognitions more possible and others less so. A logic of recognition is not a totalizing force, and people live with and despite its operations in all sorts of ways. But this never means it is inconsequential, either. Recognition is an attempt to account for the ways that we are all guided by social norms that define a range of ways of existing in the world that have been, at this moment in time, identified as intelligible to others. We can choose to make ourselves more or less recognizable and more or less normal, and this means confronting a range of social pressures and consequences that may be easier or harder to bear for a given individual based on a number of factors, including one's social standing but also one's nerve. Sometimes existing outside of the terms of social recognition may not be survivable at all.

Regimes of recognition are capable of framing and arranging reality. They are also capable of failing to acknowledge that reality. When undocumented migrants gather together in demonstrations and assert a right to appear in public and be acknowledged for the contributions they make as well as their right to exist, they challenge biometric norms that, in a contradictory fashion, suggest they do not exist or that they exist only underground, furtively surviving. Something as simple as being gathered together in public can be "a way of laying claim to the public sphere" and can expose the ways public life can depend on refusing to recognize the presence of some humans (Butler, *Notes* 41). To say that biometric thought involves social recognition includes, therefore, the possibility that those recognitions may take the form of a failure to see, including the refusal to see the exclusions that can structure public life.

Sometimes recognition is possible but unwanted and what one needs in order to live is social invisibility. When we consider the unique circumstances of individuals and the relations of power that structure their lives for better and worse, it becomes difficult to know in advance how the social recognitions engendered by biometric thought might work in all instances. We cannot assume that biometric thought will always be a hindrance or that its effects will reliably produce the same results. What is needed, then, is attention to how biometric thought governs the way individuals are recognized to exist and the constraints that come with such legibility in any given unique context rather than assume that biometric thought will affect everyone in the same way regardless of the circumstances. For example, one may have learned that a body is something that defines oneself as separate from all others and is the public face of who one is. In one instance, this may not be a tolerable reality if one exists at odds with this alignment of self to body in some way, perhaps if one does not identify with conventional cultural understandings of biological sex or a visual logic of routine sex assessment that asks what one is before asking who one is. Or, if one is powerfully attached to the emotional lives of others and dependent upon them for a sense of self, a strident separation of oneself from others may make little emotional sense.

Operations of biometric recognition are widespread and diffuse. They do not belong to a single location within culture, nor are they attributable solely to official practices supported by police powers and the force of the law. Rather, the special force it has within highly consequential moments of identity verification comes, in part, from how routinely it is re-inscribed. It sounds strange to try to mount an argument about what is missing from a passport or ID card photograph, for example, because I have learned that the passport or ID card is a representation of a complete and fully independent individual. I cannot say exactly where these ideas were learned because they are repeated so regularly and issue from so many locations. Surely this understanding depends on ideas I have encountered—of my own absolute independence—in media, in forms of individualized discipline and reward first in school and then in work, in economic and legal conditions that remind me I am responsible for myself, and so on. Nor can I say that these ideas matter only to biometric thought or more so than to other areas of thought. But biometric thought clearly leverages this way of understanding individual identity and helps me, once again, recognize the obviousness of it.

Biometric recognition does not take place in a field of knowledge that it entirely controls or that it has wholly created. Rather, its social recognitions are made more and less possible on the basis of existing social ideals and norms generally, including its own history of recognitions. Indeed, what I see in biometric thought is the uneasy and unarticulated existence of older and more advanced forms of biometric thinking simultaneously. There may not be a great paradigm shift that arrives with technological developments in the 1890s or the 1980s or with the event of 9/11, after which biometrics suddenly became operational in ways it was not before. Biometrics does not suddenly prove that you are who you really are by discovering your essence in the pattern of your iris; it remains a practice of matching one piece of evidence to another stored in a database and can assert only whatever truth this database contains. Dressing up this circular operation of institutional knowledge and its exclusive recognitions of the limits of identity with the romantically real qualities of physical bodies subjected to advanced imaging technology does not make them any less artificial, any less the product of human invention.

Demir's photograph of Kurdi's death may record the loss of his life and mark it as a tragedy. Can it document the non-present conditions that made this outcome possible, however? These conditions include the destabilization of Iraq after years of war and the too slow development of an effective government, ongoing dissent and rebellion and civil war in Syria, the persecution of Kurds, the use of chemical weapons upon citizens by the Syrian government, the failures of the UN to protect civilians in Syria, and the international community's decision to largely limit their response to dropping bombs. This image makes a tragedy recognizable, but it can do little to ask why and how such events continue to unfold daily in the Mediterranean as I write this in 2019. It largely invites viewers to recognize the tragic and impossible choice faced by the Kurdi family and to lament that so many others choose to risk their lives on smugglers' overloaded boats. Responsibility becomes entirely discrete under this logic, attached to the migrants who, through no choice of their own, have been forced from their homes and have found so few livable conditions since versing. If responsibility is to mean anything, it must be able to acknowledge the conditions that structure moments of decision and limit choice. In this image, Kurdi can come to be separated from a complex world of lived relations by the most violent of biometric recognitions that would see only his death and which would try to enact a second death upon his social existence by finding the conditions that led to his death to be inevitable. They are not. They are the effect, in part, of global actions and inaction taken frequently in the name of many like me, whose government was bombing Syria at the time of Kurdi's death.

Biometric thought imagines a world composed of borders. Its calls to authenticate identity are premised on the defining presence of borders as part of the ground for how we live in the world. People are dying as they cross borders around the globe, and these deaths will never be prevented by recognitions that insist the border is primarily a carefully ordered site of identification and entry. Can we see in the image of Kurdi that borders operate as a zone of antagonism, literalized in the form of the Aegean Sea? Far from a checkpoint or an entrance, this border has become a space of life and death. In 2015 alone, more died crossing the Mediterranean into Europe than died in New York City during the terrorist attacks in 2001. The citizens and activists in Greece who notify the coast guard of vessels in danger of capsizing attest to the truth of this antagonism in the face of Europe's failure to safeguard the lives of migrants. These activists courageously refuse to shut others out in the name of self-defense or security or their own meager resources. Those detained behind bars on the island of Lesbos recognize this antagonism too and feel the weight of a European Union torn between accepting migrants and refugees and holding them at bay. Times of crisis can produce profound expressions of humanity on all sides and can be a catalyst for thinking about how we live together in spite of lines drawn on maps, even as they reveal the lengths to which the established order will go to maintain its way of life. Or they can produce forms of abandonment that refuse to recognize that a border that routinely claims lives is not an inevitable effect of the sea but a conscious decision to allow it to exist as a space of antagonism and agony.

Recognizing the border as a site of antagonism rather than one of precise adjudication means acknowledging that border crossings are life-and-death matters, not merely exercises in following the rules and verifying documents. We do not need more dreams of regulation and order for spaces that we have designed and allowed to become deadly. The image of Kurdi proves the border is not a well-regulated point of transit. The border is not a scene of inspection and identification; it is a force greater than the sea. And if this is true, what we ought to see then is an image of a child whose demise was not a natural event but the direct and expected outcome of decisions taken to sacrifice his life.

Perhaps the most perceptive commentary on this image and what it makes visible came from someone who could be forgiven, for powerful reasons of history, had he chosen to be more reticent about migration: in the weeks that followed the publication of this photograph, a hereditary chief of the Mi'kmaq nation in Nova Scotia argued for the right to admit Syrian refugees directly into its territory (Roache). Stephen Augustine commented, "We as Mi'kmaq people still have the aboriginal right and aboriginal title to Mi'kma'ki, or the territory here in the Maritimes, and why should we be stopped from welcoming Europeans or people from another culture to come here to seek shelter?" (Roache). His call reveals a great deal about the relations of power and histories of migration that shaped North America. Augustine's comments refute a narrative that finds migration to be dangerous, unwelcome, unnatural, and economically irresponsible. Augustine refuses to participate in a biometric logic that would look at this reality and see only a portrait of the migrant. Instead, he recollects histories of migration that have reshaped the globe and destroyed so many indigenous lives in the Americas. And he asserts the Mi'kmaq's tradition of hospitality, courageously refuting isolationist responses that fear outsiders and difference and which would repeat the ways Europeans neglected and ended the lives of so many indigenous people in the Americas during colonization and the creation of empires. Human life needs a world of lived social relations and relations with the land itself, and Augustine saw that in this image. His call insisted on hospitality and on seeing more than just Kurdi's isolated body. His remarks recognize the ways one can share the resources and supports of the planet together in order to survive. He refused to see violence and loss as inevitable and natural. He refused to recognize an antagonistic relationship to the land as one that should define how we exist together.

I want to entertain a few objections to my consideration of biometric thought as a practice of social recognition. Given the time spent discussing a photograph, does this approach to biometric thought reduce everything to a question of representation and a matter of how some individuals are represented? Is the answer to develop more accurate and honest representations?

Representation is important as a category for thinking about biometric thought, because it is clear that biometrics frequently involves representing the individual via particular aspects of bodily data, including photographs. When a computer processor generates and interprets a data set that represents a scan of my fingerprint, it verifies my identity by representing me via the form of a fingerprint. But biometric thought involves more than just identity verification; it establishes and reiterates a way of thinking about the world, myself, and others that is premised on my independence from others and from the world that I rely on to exist. When biometric thought makes "me" recognizable in particular terms, it does not represent reality. Biometric thought does not reflect an essence that is present for all to see. It is a mode of recognizing what it means to speak about a self and my borders, where I begin and where I end. It organizes reality according to its regimes of veridiction, to recall Foucault's phrasing that I explored at the end of the previous chapter. Biometric thought defines the conditions that make it possible for me to recognize myself as a discrete entity bound to a physical body as its ground.

This is why I write of recognition rather than representation. Representation tends to refer to the presence of something that is made visible, perhaps in a new or novel way. I use the term *recognition* to capture what I think is a slightly different matter. Recognition does not reproduce a fully present reality. Instead, it is a mode of selectively recognizing some details and neglecting other elements that make up reality. Its operations are guided by social norms that constrain and enable what one can see or hear—or recognize—about that reality. When biometric thought recognizes that identity is written upon one's body, it guides perceptions of reality by emphasizing those parts of identity that are physical, detectable by technology, and understood to be reliably present. This recognition also relegates important parts of who one is to social invisibility, at least according to these terms of recognition. It is worth pausing here to note that powers of recognition are not absolute and may fail in some circumstances, despite their authority. Likewise, they can be forced to fail if enough people refuse to recognize their authority. This is another important feature of recognition: it is a relation of power, and as such it is open to resistance and can be tested and its effects redrawn, though perhaps not in all instances and not equally easily for all individuals. If we understand representation to involve a process of mirroring what is present, this does not effectively address the relations of power that regulate how one appears as well as who is recognized to appear.

What about all the ways in which individual encounters with biometric thought differ? Can it be said to produce uniform outcomes?

By attempting to offer a sense of the effects of biometric thought—which include the recognitions it produces and the effects of those recognitions as well as the way biometric practices directly police and regulate human movement—I do not wish to inadvertently idealize or normalize only one experience of what it may mean to confront this biometric ideal of human equivalence. Just as it suggests a common capacity for individual legibility, biometric thought produces a shared set of recognitions that guide how we recognize ourselves and others. The depth and consequence of those recognitions and how they are complemented by other relations of power matter greatly. If the capacity to be recognized by biometrics is widely shared, it does not follow that the effects of that recognition are equivalent. Some individuals will be negatively affected much more than others for reasons of power and economics, for cultural reasons, or because of the histories of truth to which biometric thought is attached, and the effects may be more or less pronounced in given circumstances—all of which might be true simply because of that axiomatic observation Eve Sedgwick made years ago: "People are different from each other" (*Epistemology* 22). We may not know in advance how effects will unfold, but I certainly agree we must remain attentive to the differences that will always arise in different circumstances and for different individuals, groups, and communities. Any serious attempt to consider the occlusions of biometric thought demands it.

For example, some may find that the ideas behind biometric thought directly affect how they live and interact with others, even though they may never leave their local communities. For others, these ideas may be backed with police force during inspection at a border. And for others still, such ideas may be part of an experience of being stateless and unable to verify identity in a manner that will be recognized and thus put them at risk of remaining permanently abandoned. For others, it may be as incidental as unlocking their cell phone. Many different modalities and relay points materialize the biometric assumptions about the ground of existence and the tremors they issue unevenly to so many adjacent concepts.

Are migration and mobility always a good thing? It may often sound as though I am, implicitly at least, making an argument for migration as much as I am offering a consideration of what biometric thought involves. What does such an argument mean for place-based movements all over the globe that express as foundational an attachment to the land or to a particular place and the history it represents? Residents of Boyle Heights in Los Angeles, for example, have begun a spirited resistance to intra-urban migration and economic displacement in the name of development and improvement. Less dramatic than the experiences of refugees fleeing war, gentrification nonetheless upends lives and scatters communities permanently, and this form of migration should not, inadvertently or on purpose, be seen as necessary or inevitable. In Boyle Heights, residents are challenging incursions into their neighborhood, knowing that it will drive rents up and force a community to disperse. The point, I think, is not to argue for human movement as natural but instead to see it as a consequence of something more than individual desire. Human movement has causes that are complex; it names an event and a set of recognitions that are neither uniformly desirable or undesirable but situated instead within complex relations of power and contexts that make each instance unique and in need of interpretation. As an approach to thinking about how biometric thought regulates and recognizes human movement as an object of regulation, the method I emphasize here is designed to identify the particular frameworks that make a given assessment of movement possible and to consider whose interests that assessment serves.

Throughout this chapter, I have examined some of the social recognitions produced by biometric thought that regulate and condition how we think about ourselves and others, about human movement as something to be constrained, about the causes that compel people to move about the globe, and about the ethical responsibilities that so rarely figure in such discussions. I want to briefly highlight the ways in which this approach is decidedly different from what we see in extant intellectual assessments of the human experience of biometric technology.

My goal here is to further distinguish biometric thought from existing ways of understanding biometrics before insisting, as I do in the next chapter, on the importance of understanding the history and durability of biometric thought.

In *Biometrics: Bodies, Technologies, Biopolitics*, Joseph Pugliese contends that biometrics is an extension of a technology of governing that sees dangerous, deviant, and undesirable individuals at the root of social problems and seeks to regulate and correct those individuals. This is a mode of governing that Foucault analyzed in terms of the development of a normalizing society that sees crime to be the result of a criminal mind, not the name given to particular acts that a society has outlawed; crime is a consequence of individual deviancy. To give a quick example of my own, consider the Russian punk band Pussy Riot performing illegally in the Cathedral of Christ the Saviour in Moscow, for which band members were found guilty of hooliganism and religious hatred. These charges silence the nature of their political protest against the church and state and their effort to illustrate how the two have been braided together in contemporary Russia. Likewise, these charges ignore their protest against a conservative culture that encourages violence against women, gay, lesbian, and transgendered individuals. By marking these women as hooligans, the Russian state ignores and invalidates their political criticism by identifying their expressions only as evidence of their naturally deviant personalities: they are hooligans who hate religion. As a strategy of governance, this method works to see individuals and to explain all activities in terms of individual pathologies rather than admit that social conditions and contexts might inform how one lives and acts.

For Pugliese, biometric technology poses the question "Who are you?" and answers it by "establishing the specificity of the subject's embodiment and her or his geopolitical status," such that "what you are—a person of colour or an asylum seeker—determines the answering of who you are" (1–2). Much like my own, this approach enables Pugliese to note that biometric technologies reveal the existence of normative categories that structure how we see individuals. These categories include "whiteness, heteronormativity, ableism, class and geopolitics," and they "enable enfranchisement for some and disenfranchisement for others" depending on how closely they can conform to a given normative ideal. Biometrics is largely a tool, in his account, that confirms the settled existence of long-standing regimes of power.

Pugliese sees biometrics as powerful and unwanted, then, an oppressive technology that makes many lives less livable. There is no room in his account to see that this is a relation of power and as such it is subject to revision and alteration. Nor is there room to consider that biometric relations of power might be anything other than oppressive or that they could be productive in the sense that they frame the world and bring it into focus in particular ways. Instead, its power appears to be total, consistently oppressive, and unstoppable. If biometrics is a tool, it can only be wielded in one way and produce one effect, apparently. I am not certain this is true. I am quite certain that assuming it to be true does more harm than good. If we agree it is important to think about the ways in which biometrics is enmeshed in social relations of power, it is all the more important to think about how biometric operations may make possible opportunities to challenge its function and reach.

For Pugliese, biometrics does not identify individuals but instead discovers them to be expressions of normative types. His approach rightly recognizes the presence of societies defined by discrimination, but his argument largely ignores the social existence of biometrics as a mechanism of identification and the recognitions it makes regarding the nature of individual identity. A society grounded in recognition—of oneself by another, or by forces outside oneself—may well produce discrimination, but that is not all it produces. For example, Pugliese largely assumes a notion of possessive individualism, of discrete individuals who are seen to belong to a race or a gender, without asking what makes a body possible or what sustains and supports embodied life. Individuals are a given, in his argument, whereas I contend that the very idea of the individual as something separate from others and dangerously independent from a social world and from the earth that sustains it is a recognition that biometric thought helps to make possible and obvious. There is a social life of biometric thought that arranges and makes reality legible, I contend, and this includes a number of social recognitions that govern how we think and act in the world.

By taking this approach, I am better able to address the ambitions, desires, and indeed promises that make biometrics appealing. If biometrics were only an oppressive practice of social control, it would likely never take hold in free and open societies. Biometric thought offers a number of obvious recognitions around security, vulnerability, and the pressing need to regulate global mobility that can be socially appealing, but it also makes possible recognitions that define the very essence of identity: the mental concept that one is a unique and discrete individual is a vital counterbalance at a time when capitalism makes a great many feel less unique and more replaceable than ever before.

What I find least persuasive in Pugliese's account is the assumption that the possibility of a given technology to do something unwelcome means it will do that unwelcome thing at some point and that this ought to be understood as the essence of the technology itself. Biometrics could violate privacy by recording a medical condition (98), or it could be used to racially profile individuals (75), or it could be prejudiced against individuals who are embodied in ways that biometric technology cannot effectively acknowledge and who thus might "literally fail to be counted as human subjects" (89). A great many effects of biometric thought are presently shaping the globe and undoing the ground of our shared social existence, and these become less comprehensible under an approach that worries over what biometric technology might do. Rather than contemplate what the technology might do, particularly when it is used for ends other than verifying identity, Shoshana Amielle Magnet considers what the technology accomplishes when it fails to work as it claims.

In *When Biometrics Fail*, she contends that biometric practices of identification are not nearly as effective as we believe or have been led to believe by companies producing such technology and by popular representations of it. Failures of the technology include mechanical deficiencies, failures to meet basic standards of objectivity and neutrality in application, and the failure to adequately conceive of the variety of human subjects who will be registered by the technology, all of which necessarily call into question the claims and aspirations of biometrics. Yet despite persistent failures, she notes, biometric technologies still accomplish a great deal for state and commercial actors whose interests are tied to contemporary cultures of security and fear. I share with Magnet the idea that biometrics is a productive technology rather than an oppressive one, an idea that she renders as the capacity of biometrics to fail and still do things such as "produce new forms of identity, including unbiometrifiable bodies that cannot be recognized by these new identification technologies, a subject identity that has profound implications for individuals' ability to work, to collect benefits, and to travel across borders" (5).

Additionally, she notes the technology may fail even when it functions in a manner that is technically correct. Biometrics does real damage to vulnerable people and groups, to the fabric of democracy, and to the possibility of understanding the bodies and identities these technologies are supposed to protect. In this sense biometric technologies fail even when they succeed (2–3). Her attention to error reminds me of something Foucault once noted as a feature of Georges Canguilhem's philosophical account of the history of science: "the opposition of the true and the false, the values that are attributed to the one or the other, the power effects that different societies and different institutions link to that division—all this may be nothing but the most belated response to that possibility of error inherent in life" ("Life" 476). If evolutionary biology teaches us that life is animated by a series of errors and uneven corrections, then we can see that the very idea of *correctness* is a remarkable invention for a species that "from the depths of its origin, bore the potential for error within itself" ("Life" 476–477). We should not be surprised, then, that failure might prove to be every bit as productive as success for a human reality that may respond more to error than it ever has to social distinctions between truth and falsity.

Magnet's analysis belongs to a history of evolving technology more than to a history of the present that would establish the parameters of what we know to be true, however. Indeed, this is the biggest difference between my own approach and Magnet's. I am tracking how biometric thought produces a recognition of a set of ideas as true or obvious and how these are conditioned by a history of biometric thought that encourages understandings of oneself and others. Magnet tends to see biometrics as something utterly new, limiting her analysis to recent advanced biometric technology, which is something that she interprets to be largely without precedent and attached to American matters, such as security concerns since 9/11 and efforts to regulate social assistance by making recipients enroll in biometric identity verification.

Her argument contemplates the possibility that mechanical and procedural neutrality might yet arrive if only these failures can be addressed and overcome by the engineers, scientists, and technicians developing biometrics. Yet Magnet's emphasis on the profit motive that drives much of this industry means that the argument cannot quite dismiss a more cynical response to the failures of biometrics from those within the industry whose inaction would seem to say, "Who cares? The present deficiencies are not hurting business and if the technology does not have to be effective to appear effective, all the better!" Magnet does not suggest this conclusion, but her argument leads us to imagine that the political expediency of biometrics combined with an influential industry means that the technology does not have to be effective at all. She posits an industry that acts largely without regard for the failures it makes. By contrast, I argue for the existence of biometric thought in order to note that the effects of biometrics are not limited to the interests of those involved in creating technical procedures of identity recognition. There is a "doing" here that needs to be understood as something that happens socially rather than just individually or according to an agreed-upon set of priorities such as profit or security. Biometric thought makes a whole host of social recognitions possible, and its effects reach much more widely than the operation of biometric procedures and devices or the aspirations of CEOs. More, these are not necessarily only associated with advanced biometric technology. Present biometric practices are possible because recent technological achievements have been grafted onto existing social forms, existing regimes of truth, and thus I contend that biometrics has a history as a set of operations and as a mode of thought that informs those operations.

In *The Politics of Humanitarian Technology* (2015), Katja Lindskov Jacobsen examines the use of biometrics in humanitarian contexts such as the registration of refugees. She notes that such activities belong to a history of experimentation upon subjects who lack the political clout or sufficient information to resist or, I would add, to consent to such treatment (59). I make mention of consent because I do not want to simply see every instance of biometric technology as inherently or uniformly unwelcome and harmful.

Experimenting upon vulnerable populations has a long history. Jacobsen examines humanitarian vaccination programs, in which the desire "to deliver medical assistance to populations in need may very well be based on good intentions" but can nonetheless overlook the "conditions that allow for humanitarian subjects to be regarded as suitable for exposure to experimental uses of new technologies" (127). One might recall medical experiments such as the Tuskegee Syphilis Experiment that began during the Depression in 1932 in rural Alabama in which the

U.S. Public Health Service studied the course of the disease in untreated African American men while claiming to administer free health care. This history of experimentation also includes developments in government and education. Within my own discipline of literary studies, much of the purpose and many of the tactics of aesthetic education was worked out as part of the civilizing mission of the British empire and its educational measures in British East India in the nineteenth century, as Guari Viswanathan has noted. In *Decolonizing the Mind*, Ngũgĩ wa Thiong'o famously relates his experience of cultural imperialism growing up in Kenya that included petty schoolyard games in which children passed a button to anyone caught speaking a language other than English and were subsequently punished for speaking tribal languages. There is no shortage of great violence— direct or cultural or, more commonly, both at once—that has been carried out in the names of assistance and development.

Lamenting the experimental deployment of biometrics, Jacobsen notes that such trials are given over to problems of data storage, privacy, and limitations of the use and efficacy of the technology:

> It is impossible to acquire exact data about the number of false matches in this context, but a simple calculation suggests that for an enrolment population of two million (which is a low estimate), an error rate of 3 per cent (Moore 2003b) translates into 60,000 false matches. This is, of course, not an accurate representation of the number of Afghan refugees that may have been falsely matched during this endeavor. Indeed, I do not want to suggest that such a high number of false matches went undetected in this UNHCR operation. The point to stress is simply that this risk of technical failures (particularly under experimental conditions) being translated into humanitarian failures may have been of a magnitude that deserves attention. (64)

Jacobsen is concerned with hypothetical failures such as the "the various potentialities for negative implications for the safety of Syrian refugees—negative potentialities that humanitarian representations of the ability of biometrics to improve humanitarian efforts in the context of the Syrian refugee crisis, may largely overlook" (79). Jacobsen's approach is largely empirical, even if it is interested in what might yet develop, and is thus limited to discussing only what biometric practices can make plainly present via their operations. My approach, by contrast, is defined by its break with a strict empiricism, in the sense that my focus on recognition implies attention beyond the discrete real-world matching function of biometric verification to encompass the social norms regarding identity that subtend and inform its operations as well as attention to how such norms function more widely to regulate how individuals exist as if in isolation from one another.

The greatest accomplishment of biometric thought may be its ability to identify existence in such a way as to remove all context and disconnect individuals from one another and from the world around them. So long as we are looking only at what biometrics makes visible, reveals, and unveils in its operations, we will remain unable to account for this powerful effect. Biometric thought constitutes an accomplished practice of normalization in which individuals are separated from one another and made to appear as if in a vacuum. Unacknowledged, the social basis for life appears either impossible to comprehend or, at best, optional and unnecessary. Biometric thought loses much of the world and estranges us from the human communities, as well as our ties to the earth, that make life possible and from the shared responsibilities we have for events in the world. How did this come to be and what forces have made this present possible?

5 · A GENEALOGY
OF BIOMETRICS

In his study of techniques of identification, historian Edward Higgs suggests that our present era is distinguished by the disappearance of an identity one can assert on one's own terms. One must "claim it from institutions" and from "official registries and databases" (205). The idea that identity is something outside of oneself is not new, however, at least if we recall Hegel's fundamental insight that existence depends upon being recognized by others. And while this might sound benign when addressed to bureaucracy, we can recall Kiese Laymon's experiences of the ordinary violence of racism that is likewise part of what it means to experience an identity that is not fully of one's own determination. This chapter traces a brief history of biometric technology as a mode of recognition that seeks to grant identity and treat individuals as if they need verification. My premise is that biometric thought is much older than advanced biometric technology might lead us to believe and that its effects may not be as new as some think. While I cannot offer an exhaustive or even substantial genealogy of what preceded our biometric present here, this chapter suggests where it has been particularly concentrated and how it has altered over time. This includes the development, removal, and reinstitution of international rules for identifying individuals and regulating mobility and their shifting goals, as well as parallel technologies of body-based identification such as fingerprinting, the rise of the science of biometry, and physiognomy. Ultimately, the chapter explores one of the more discredited antecedents of modern biometrics at length—the popular idea of inferring identity by reading the face as if it were a book—in order to remember that a key part of the history of biometrics involves contests over what this technology does and can claim to do, rather than a record of its settled and punctual accomplishments. Such contests document the ways in which a technology can have multiple social lives and be enlisted for a variety of ends and goals.

The passport has long been one of our most common physical manifestations of biometric thought. The history of the passport is less a history of travel or movement than a history of record keeping and the institution of identity as something that must be officially verified by others. Passports were the result of uneven and inconsistent efforts to regulate and register movement and to keep records about such matters. Though they are an invention of the nineteenth and twentieth centuries, they have their origin in licenses granted by monarchs to their subjects as well as letters of safe passage given, for example, to representatives of a foreign ruler coming to negotiate a particular matter upon entry to the territory. Visas are a modern equivalent to a safe-passage document, issued as they are for a traveler by the country they are visiting, though they are now designed more to protect the nation than the individual traveler.

As smart as a biometrically encoded passport might be or yet become, it remains a document enmeshed in larger structures of knowing and information tracking. Mark Salter contends that the passport is an insecure document that can represent identity only because it references "other insecure documents" in a chain of institutional knowledge keeping in which identity is always a game of matching one record to another (84). Yet the rules and goals associated with passports have shifted considerably over time. This is apparent with even a cursory glance at just how different some measures were three hundred years ago. A 1719 Russian edict required subjects to possess a pass granting the right to move from one town to another (Torpey 19). At the same time in France, the term *foreigner* designated someone from outside a local region (20), suggesting that the threshold at which another becomes strange has moved over time and that movement itself was largely uncommon. Registration and identification were often undertaken in order to restrict movement rather than to create the globally connected society we have now. Over the course of the nineteenth century in Europe, the lower classes gained more and more freedom of movement. This was not an inevitable or unregulated shift, however. It coincided with the advent of economic liberalism and a free market of labor that was designed to destroy an older guild system that regulated professions and wages (89). The shift toward free internal movement was complemented by the rapid increase in movement beyond the borders of the nation, particularly by a middle class that now possessed the freedom and means to travel the globe in much the same way the aristocracy had, but at a scale previously unmatched thanks to the creation of railroads and steamships. Part of what this period shows so clearly are the ways that human movement was regulated by economic conditions.

The regulation of movement was routinely contested, the historian Martin Lloyd notes. French laws that controlled "the movement of peasants from the country to the towns . . . to prevent capital and skills from moving out of specific geographical areas" were abolished by the revolutionary government in 1792 (65). This led to greater freedom of movement not only for French citizens but also noncitizens within the nation. This would quickly change under Napoleon, who sought to register foreign nationals as potential enemies of the state. Yet if war often prompted European nations to closely monitor the arrival and residency of strangers, the desire to record identity and track movement often waned in peacetime (Higgs 208). "Even in the twentieth century," Higgs notes, "the British State showed itself unable to sustain national registration systems and ID cards outside wartime, and was unwilling to use the new biometrics system of fingerprinting to identify citizens as opposed to criminals" (208). While often understood to be an age of increasing governmental regulation across Europe, the mid-nineteenth century saw passports abolished by Norway, Sweden, Italy, and Portugal. In Latin America during the same period, "countries such as Venezuela, Uruguay, Mexico, Ecuador, Bolivia, and Peru had constitutions in which the right to travel freely without a passport was clearly stated and extended to all foreigners" (Lloyd 115). We should not think that our present arrangement of tightly controlled movement and pervasive registration and identification of migrants and travelers is the only world possible or the inevitable march of the progress of our times. History shows that other arrangements have been viable and that pre-existing forms of biometric thought have been renounced by those alternatives.

As formal documents, passports can be traced to a European origin largely because European imperialism and colonization made it possible to export and enforce Europe's model of identification into the distant corners of its empire. Yet the authority of passports was slow to emerge because of their haphazard history as documents that were designed to accomplish distinct goals. Until the middle of the nineteenth century, passports were not issued from single authorities within nations and did not possess a consistent form. This was as true for the federated states of Germany as it was for the United States, where local authorities often issued travel documents with requirements that varied quite widely. Throughout this period, passports were entwined with concerns around immigration, anxieties over so-called foreign influence, and caring for the poor. Eighteenth-century concerns with individuals leaving the territory were replaced by concerns with individuals arriving in the territory. What remained constant were the regulatory ambitions of the passport, even if its function could shift from ensuring mobility to restricting it, as circumstances required, and often worked in a decidedly unsystematic fashion. As scholar John Torpey notes, the 1868 Burlingame Treaty between the United States and China permitted the free immigration of Chinese citizens to the United States without any corresponding right to citizenship, making these migrants "early versions of what came to be known as 'guest workers'" (96). Yet only fifteen years later, the United States passed the Chinese Exclusion Act, which sought to bar Chinese immigration and eventually created an entire apparatus of registration and identification that controlled the movement and rights of re-entry for Chinese immigrants who had arrived under the treaty. For these individuals, passports and associated regimes of identification became a means of recognizing children born in the United States as citizens while aggressively constraining the mobility of their parents.

Similarly, in 1982, South African courts asserted that passports are a privilege, not a right, and the apartheid state "could revoke any passport without cause or appeal" (162). This measure ensured that passports, like other identity documents, would remain a means of regulating the different freedom of movement allocated by the South African state to those identified according to the legal categories of Bantu, colored, white, and Indian. The Internal Security Act of 1950 in the United States regulated the movement of citizens and ideas by denying passports to suspected members of the Communist Party. More recently, Canada joined a number of other nations and created a law in 2014 that revokes the passports of individuals suspected of traveling out of the country with the intention of joining the Islamic State. Passports function to constrain movement at least as readily as they enable it. They do a great deal more than simply affirm the identity and nationality of the bearer. As an expression of biometric thought, they ground individual identity within a regulatory schema that seeks to regulate mobility and is supported by police powers to restrict human movement when deemed necessary. The world imagined by the passport is one in which permission must be sought for movement, even if we tend to represent that little booklet as an enabling condition that makes travel possible.

The very idea that a passport is a document of individual identity has not always been true. To illustrate this point, Lloyd shares the story of King Louis XVI and Marie Antoinette nearly escaping France after the French Revolution in 1789. During their journey, the deposed king and queen had their carriage searched and papers inspected, yet "the royal fugitives had less to fear than most since, according to their papers, they were unnamed servants travelling for their foreign masters, the Korffs" (Lloyd 63). Two hundred years earlier, French merchants used passports "issued to parties of travelers together with their goods" (Groebner 175). Who and what constitute someone worth identifying has changed over time with the gradual expansion of dignity and rights to all regardless of class, gender, and race. The idea that biometric thought recognizes all bodies in the same way has not always been the case, if it is even true at present.

As a document of identification, passports have long contained descriptions of the individual they identified. Concerned that the king might again try to flee the country, France quickly mandated that passports include name, age, description, and the parish to which the holder belonged (Torpey 27). Less important than the given details are the ways that a biometric logic establishes the ground for identity and how, over time, it ceases to acknowledge the existence of any other possible ways of accounting for who someone is.

The addition of a photograph to the passport is not a great or remarkable invention in that regard, either. Instead, it is a continuation of a practice of selectively identifying an individual according to some details and not others, largely based on appearance. The passport began to include photographs first in the United States in 1914, at the outbreak of World War I. After the war, a League of Nations conference in 1920 established the shared criteria for the present booklet-style international passport that contains a photograph and identifying details. As Lloyd notes, the re-emergence of strict travel restrictions following the nineteenth century's openness resonates with the present moment and the way that concerns with terrorism have been used to suspend civil liberties and introduce advanced biometrics: "The reality of war allowed governments to introduce into law restrictive measures of a ferocity such as would never have been permitted in peacetime, while cynically qualifying them with the assurance that they would be lifted once hostilities ceased" (119). The process of arriving at an international standard was far from seamless, however; nations debated matters such as what norms determined the front and back of a booklet, the nature of the security measures, and the fees that would be charged (122–130). The emergence of a common international method of determining citizenship and identity and regulating human mobility was a coordinated endeavor, though it failed to solve the crisis that largely made it necessary: substantial numbers of refugees, particularly from Russia, displaced during World War I and the Russian Revolution of 1917. "Contemporary estimates showed 600,000 refugees in Poland, a half a million in Germany, 70,000 in the Far East, and 150,000 in Turkey" (132). Russia refused to provide documents to citizens no longer within its territory. Many countries were unwilling to accept refugees who lacked documentation because they would become stuck within the host country and be unable to move internationally unless they were granted documents reserved for citizens, something host nations were not ready to do. The eventual response developed by the League of Nations was the Nansen passport, named after its creator, Fridtjof Nansen. Unique in the world, the Nansen passport was attached to no country and guaranteed no citizenship, but it enabled movement and became the foundation for the United Nations' capacity to issue documents to stateless individuals and provide them with protections alongside the identification needed to obtain work permits and residency permits.

Identification had become something that had to be officially possible, even in instances when it was officially impossible. Identity had become reliably documented and documentable, attached to the unique individual and a necessary feature of international movement. This was a welcome innovation, appreciated by millions of displaced Russians and Europeans, and is similar to the way advanced biometrics can make it possible for refugees to return as documented citizens with rights and protections. The effects of biometric identification are never uniform and never isolated to its most direct purposes and humanitarian ambitions, however, and its history is not limited to passports and the select forms of documented existence they recognize.

Indeed, the practice of identifying and recording the identity of those charged and convicted of crimes helped to define biometric thought at least as much as the passport. Over the course of the nineteenth century, jurists, police, politicians, and those involved in studying the sociology, anthropology, and psychology of those who had committed crimes developed the theory, in many ways still with us today, of the criminal as a type of person. While some might be occasional criminals, many more were hereditary criminals who could not be reformed, argued the Italian criminologist Cesare Lombroso in the late nineteenth century. Acting on this powerful assumption, criminology adopted many of the pseudoscientific intellectual moves of phrenology that inferred moral and intellectual capacity by measuring the physical shape of the head. Criminology began as a method of measuring features such as facial asymmetry, the set of the eyes, the jawline, the shape of the nose, and the length of the arms. The goal of such measurements was twofold: first, they would reveal and prove the hidden nature of an uncaught criminal or even an as-yet-inactive born criminal on the basis of telltale physical features; second, these measurements could effectively track individuals who were repeat offenders even if they used multiple aliases and had multiple forms of identification. Criminology confronted a real social problem of identifying and tracking mobile individuals charged with crimes. Its practices reflected new assumptions about recidivism and the genetic basis for crime that ignored social circumstances, structural inequality within a society, and social imbalances between what is and is not considered a crime and the severity of the punishment that it merits. When one sees only faces in the context of criminal activity, much has been lost.

Identifying criminals is not nearly as empirical a process as it sounds. Throughout the 1980s and 1990s, for example, federal law in the United States established significant differences in the penalties associated with drug possession. The distribution of five grams of crack cocaine resulted in five-year mandatory sentences, while it took five hundred grams of powdered cocaine to merit the same penalty (ACLU). Although the substances are pharmacologically identical, the users of crack are predominantly African American and poor, while the users of powdered cocaine are largely white and wealthy. Such discrepancies are a direct result of a criminology that seized upon the criminal, not the crime, and sought to identify this person individually, anthropologically, and visually. Such a practice relies on a politics of recognition that can see only some details as relevant. This practice continues to explain why some may be quick to identify violent crime and slow to recognize great acts of harm that have become legible, instead, as acceptable ways of conducting business, investment, and trade. Some individuals look like criminals, while others look like respectable members of society. This is not an accident; it is a legacy of Lombroso's efforts to understand that crime has a very particular human face.

I am reminded of a striking performance art project entitled *Captured: People in Prison Drawing People Who Should Be*. The portraits created by incarcerated individuals include numerous CEOs whose companies have recklessly prioritized profit over all other considerations, and beneath each individual is a prison ID

number. I cannot help but marvel at the considerable talents of so many impris-
oned artists and the humanity of the images they produced as part of this project. *Captured* reminds viewers that some people can openly do harm to others
because we have sanctioned these particular violations as permissible according
to the laws that govern business, laws that are likely to levy fines that can be readily absorbed as part of the cost of doing business when illegal activities are pros-
ecuted. What is jarring about the project is that it exposes how selectively criminal harm is given a human face when it involves matters such as pollution,
exploitation, and the theft of land and resources, which tend not be recognized
as crimes at all, in part because they are conducted as if by corporations.

Our biometric present is perhaps impossible without the efforts of Paris police
officer Alphonse Bertillon, who developed a sophisticated practical manifestation
of Lombroso's desire to identify the criminal at the turn of the twentieth century.
Bertillon drew on existing ideas about reading faces to catalog the features of
arrested individuals. This was a science of isolation, as the scholar Jonathan Finn
notes, that understood identity to be an essence captured by so many telling signs
on the criminal's body. Individuals were photographed and facial features were
measured. Data was recorded and files created and sorted for future identification. Bertillon's system of record keeping produced identity as a physical property of the person and something that had to be removed from any and all other
contexts. Appearance was not merely representative of who one was; it was expressive of that fact and could betray the truth of one's essence. Identity belonged to
the body first and foremost, and it would betray the corrupt soul of the offender.

Bertillonage, as the practice was called, briefly took hold in Europe and North
America. Yet in the early decades of the twentieth century, another new measure
of biometrics had begun to promise a much more streamlined mode of physical
identification: fingerprinting. Like Bertillonage, fingerprinting depended on the
newly accessible technology of photography, but in this case, fingerprinting could
reproduce evidence from a crime scene in order to compare it to a fingerprint
taken from a suspect. As important as the technical measurement of fingerprint
lines and ridges was, it was the international coordination of practices, standards,
and databases that made fingerprinting so powerful as a mode of identification.
It promised not only reliable identification of individuals by a given police force
but also possible identification of any individual from any jurisdiction readily and
with minimal difficulty, compared to the alternatives. Identity was something that
could, potentially, be recognized in the same manner, using the same standards,
anywhere in the world.

As a concept, fingerprinting reinforced the idea that the physical body is the
ground for identity. It isolated identity from dependent social relationships, culture, and a worldly existence. The idea that one's existence is defined by the marks
one leaves in the world reinforced empiricist concepts of identity—that identity
will be determined by its discrete physical presence in the world—and insisted

that identity arrives within practices of recognition and interpretation. As the scholar Simon Cole notes, this technology was useful for more than just tracking criminals. The possibility of recognizing individuals via fingerprints was particularly well suited to the colonies where, in the case of India, British officials struggled to differentiate "one Indian from another" (65). Fingerprinting not only was useful for tracking the movement of individuals, then, but also was prized as a technique of identification suited for the presumptions and racism that defined empires and colonies. As a means of establishing identity by interpreting unique physical features, fingerprinting did little to disrupt existing discriminations based on race and class, despite its potential to do so via an objective scientific method. Perhaps this is because it did not invite a widespread discussion regarding what it means to attach identity to the visible features of the body and the ways in which that presumption has been consistently used against individuals of particular races and ethnicities. Such failures to contemplate the differential ways in which bodies have been made visible and what can and cannot be recognized about individuals and races on the basis of that visibility have tended to make white privilege a foundational feature of the development of the early science of biometrics. Fingerprinting had the capacity to not see race, even if it failed to realize this potential during this period, and this universalizing capacity belongs to a tradition of biometric thought that had become entrenched in the second half of the nineteenth century that sought to identify shared human features common to all.

When Charles Darwin wrote *The Expression of the Emotions in Man and Animals* (1872), he immediately recognized that he was contributing to a genealogy of biometric thought (14–16), noting the importance of earlier work by Sir Charles Bell and Johann Caspar Lavater, whose contributions at the turn of the nineteenth century I will consider shortly because their contributions lay the foundation of this genealogy of modern biometric assumptions and techniques. Darwin saw his own work as far more systematic and grounded in evidence drawn from across a variety of cultures than these earlier works (25). One of the functions of Darwin's analysis was to calculate the extent to which the face and facial expressions could be used as a means of establishing an evolutionary continuum from animals to humans such that "man is derived from some lower animal form" (334). The significance of his study of the animal origin of human expressions "contradicted the accepted theory of the time, viz., that . . . men and animals were qualitatively distinct" and was part of an effort to further establish as viable a theory of evolution (Chevalier-Skolnikoff 13). Darwin's claims likewise established a common human heritage that challenged those who thought it impossible that all races "descended from a single parent-stock" (Darwin 329). Those who have inherited and popularized this universalizing perspective on biometric thought tend to forget these contexts entirely in the interest of a different goal: producing an idea of the face and human body generally as a legible object that can be read for the psychological depth it offers.

Such a universalizing theory of face reading has been consistently popular. Psychologist Paul Ekman, a student of Silvan Tomkins, has firmly established the psychological foundations for studying faces as legible objects, carefully recording consistent expressions that recur across cultures, identifying micro-expressions that can be the clue to staged expressions rather than involuntary ones, as well as techniques to detect such briefly visible features. The function of such research can lead in a number of directions, especially toward instances when detecting lies may be especially important. Ekman has worked to popularize this research with the idea that becoming attentive to the faces of others means learning the truth of another in the expressions of one's face and that this may lead us to "be better able to deal with people in a variety of situations and to manage our own emotional responses to their feelings" (*Emotions* xx). Depending on how it is deployed, such advice to try to understand oneself and others better may be entirely helpful. Of course, this sort of advice can also sometimes suggest that all conflicts are interpersonal, that understanding alone solves power imbalances, and that talking it out soothes experiences of oppression, all the while leaving structures that produce inequality and conflict unchanged and even unnoticed. A universalizing biometric thought may not necessarily lead to such ends, but it can end up explaining structural inequalities in individual and personal terms.

Ekman's research can be captivating, luring the reader with the promise of being able to detect, with special training, the signs of "emotional leakage" that betray "how a person feels even when the person attempts to conceal that information" ("Darwin" 207–208). He offers readers and scholars the promise of uncovering a suppressed truth, which has been fictionalized in television shows such as *Lie to Me* (2009) and *Bull* (2016). These narratives dramatize scholarship that finds the figure of the face to be especially useful for criminal justice matters in which individuals might lie or might be perceived to lie by juries despite being truthful (Black, Porter, et al 51). I cannot help but wonder how often one notices that these popular renditions of face reading stage a fundamental instability for the viewer: the truth of these insights is always delivered by actors; thus, audiences encounter an entirely simulated environment as if it were real, prompting one to wonder to what extent any of us have learned to read faces, at least somewhat, from actors trained to produce, on cue, legible expressions of emotional and psychic depth. Context matters, and the allure of reading faces is always, in part, an allure that depends on sacrificing context and eliminating complexity by seeing only an isolated face. My complaint is not that this science is false but that it can produce knowledge without necessarily examining or exploring the conditions that structure and drive its discoveries, much like the viewer whose detection skills involve seeing the truth in a carefully staged and presented scene. It strikes me that the universality of human expression should not be the basis for universalizing and naturalizing what one wishes to insist will be found or discovered in the expressions of others.

Consider, for example, present psychological studies of an unfamiliar face that asks the viewer to gauge "trustworthiness" according to "a number of structural features of the face" (Black, Porter, et al 50). In each instance, looking at the face is a means of seeing much more than a face. These are operations structured by desire and informed by presumption, including the idea of the face as a legible guide to some unknown qualities. Looking at another is structured according to a set of social presumptions about what a face reveals and can be said to make visible, and these assumptions clearly have a history and a set of purposes in mind. Darwin sought to articulate a line of descent from nonhuman animals to animals by reading faces. And Lavater saw in face reading an opportunity to parse ethnic and racial differences at a time of growing class anxiety and social diversity that made proper social relations in densely populated urban environments so difficult to parse. Assessing the expressions of others relies on and reproduces social effects that, in each of these cases, involve establishing an idea about what one sees when a face is brought into view by determining the context for that encounter (the danger of strangers, evolution, and vulnerable social hierarchies).

Reading faces has never been just about assessing information and determining its meaning. The truth that biometric assessment and apprehension produces is greater than what can be found in the lines of a face. As soon as one reads another, one does so with a desire to know something or simply to know rather than confront the possibility of not knowing anything about another. The decision to read the face and to insist on its legibility is always a way of insisting on bringing some things into view, establishing codes of subjects and objects, and deciding matters about how one exists and the status of oneself and others.

If one is heartened to discover in Darwin an argument that could contest racist theories of descent, the potential for biometric thought to map nicely onto fantasies of racial inferiority and superiority did not disappear with his reflections on human and animal emotion. In 1903, the founder of Draper's Biometric Laboratory at University College London, mathematician Karl Pearson, sought to understand genetics and evolutionary theory via a statistical analysis of life: literally bio-metrics. Pearson's work is not attuned to individual differences but instead to the characteristics of a nation or a race. His work shows that practices of assessing identity on the basis of physical and inherited traits reflected a science of racism and imperialism before anything else:

> If you bring the white man into contact with the black, you too often suspend the very process of natural selection on which the evolution of a higher type depends. You get superior and inferior races living on the same soil, and that coexistence is demoralizing for both. They naturally sink into the position of master and servant, if not admittedly or covertly into that of slave-owner and slave. Frequently they intercross, and if the bad stock be raised the good is lowered. Even in the case of Eurasians, of whom I have met mentally and physically fine specimens, I have felt

how much better they would have been had they been pure Asiatics or pure Europeans. (22)

Such justifications for imperialism and colonialism are well known, as are the ways
in which scientific thought can be used to make racism appear objective, neutral,
and necessary. Pugliese sees in this history the beginning of the race-based thinking that has been particularly adept at "identifying, classifying, and governing 'suspect' populations," which continues to this day (54) and is manifest in instances
of racial profiling and the uneven expectations of biometric data from those travelers deemed much more dangerous than others. Additionally, the unpromising
genesis of formal biometrics reveals a paradoxical awareness that identification is
a matter of assessing visible differences at the same time that it is something else
entirely. Pearson can claim that identification is a visual matter, but he clearly
shows that identification depends not on features of the person but on modes of
thought and well-developed practices of social recognition that encourage us to
see the world and individual human beings in particular ways. His calls for colonization and an outright race war in the name of "the survival of the fitter race"
(26) repugnantly remind readers that the social basis for how one thinks about
the world and the reasons individuals move about it can change and have changed.
Technologies that have sought to make the ephemeral nature of identity visible
in the body have long been engaged in acts of recognition that are premised on
not seeing human dignity, premised on selective recognitions that involve not seeing all the ways in which our physical selves depend upon so much more than
just flesh and blood. That many still justify restricting the mobility of some and
not others—and do so via claims regarding what is necessary or deserved or simply "right"—suggests that some histories of thought may be capable of impressive acts of recidivism if they manage to change their appearance just enough to
be re-admitted into civil society.

I have only scratched the surface of the histories of identification that have
shaped biometric thought and its operations. It is a not a linear history in which
practices of identification become increasingly regulated or objectively scientific.
Practices of identification have not been consistently driven by any single set of
desires, authorities, or technologies. Strategies of identification issue from government, police, science, and culture. Lines of trajectory appear and disappear
unpredictably, with older modes of identification suddenly returning to promise
again what they never quite delivered. As Higgs notes of histories of identification that have tended to focus only on documents such as passports, there are
many more places to think about how formal mechanisms of identification have
led to our present reality. He considers the development of techniques to identify the deceased "so that kin and friends can mourn their dead" (7), for example,
even if this is a less obvious area in which to reflect on the social life of technologies
of identification. One might pursue another direction and wonder about the

emotional and affective histories to be written about strategies of identification. Identification has often been construed as an unwelcome operation understood to involve some loss of a romanticized intimacy that must have existed before bureaucratic modes of identification took hold of the question "Who's there?" If there is such a thing as an authentic self, it is in part a product of the official measures of identification against which such an ideal is always defined. An authentic and intimate self is only conceivable to the extent that I have an operational understanding of my inauthentic self. In either case, identity is less something at the root of my being or my essence than it is the product of social recognitions of selfhood and modes of identification that involve others. I may be wary of a romantic notion of one's real and true identity, as if that could exist in itself rather than be yet another product of a particular regime of veridiction, which is to say a particular mode of producing and understanding what reality is and what version of recognizing that reality gets to be named the truth (see chapter 3). I know the importance of such lived and felt realities, but to imagine that I fully decide and control such matters seems hopelessly forgetful of the human social environment and the relations of power that condition every regime of veridiction. Saying that we are encouraged to make particular recognitions regarding what is true is not the same as saying that the present way is the only way of thinking about identity. Perhaps other ways are possible and may even be archived within biometric thought and its history.

With this in mind, I want to consider an older and surprisingly durable form of biometrics, that of physiognomy. As scholar Sharrona Pearl notes, the popularity of the practice of physiognomy has waxed and waned since antiquity (2), and biometric thought owes a great deal to the explosion of cultural interest in an elaborate and pseudoscientific practice of face reading following the publication of Johann Caspar Lavater's *Physiognomische Fragmente* in 1775. The Swiss pastor's return to notions of physiognomy systematically developed in antiquity struck a nerve with the European reading public, and his work was quickly translated into numerous editions across the continent, with English editions starting in 1785, including translations by Mary Wollstonecraft, Thomas Holcroft, and George Grenville and illustrated by the poet and artist William Blake, among others. Physiognomy treats visibility as a reassuring confirmation of intelligibility against the frustratingly everyday prospect of simply knowing very little about the many others who populate increasingly global and densely packed European metropoles in this period. Cities were much less stratified according to hierarchies of class and social standing than they once had been; thus, it became all the more important in the context of burgeoning middle-class ideas of propriety to discern the character and worth of another person, and the prospect of doing so visually was especially appealing. Physiognomy had found its time and place. A properly trained physiognomist could discern vital insights regarding psychology and moral character from the legible features visible on the face of so many strangers.

Embraced like never before, physiognomy and its biometric logic led Europeans to study countenances in novels of sensibility, dwell on the pleading faces of the urban poor, grow fascinated by visible differences in race in natural history, and develop cultural narratives of degeneracy in philosophy and science, all evidenced by an exploding physiognomic literature that ranged from the.luxurious gift book to the *Pocket Lavater* (1817), an inexpensive guide to reading faces designed to fit in a breast pocket and be carried into the streets of London, Paris, or Venice. If we are tempted to think of mass identification as something associated with government control and immigration and borders, the practice of reading faces demonstrates that biometric thought and technology have long been allied with less formal operations. It is a practice rooted in a widespread social project of defining the terms within which individuals can be recognized as such.

Physiognomy is older than Lavater's decisive popularization of the practice, dating back to antiquity with Aristotle's considered attention to the subject and more recently with disciples of Descartes. Groebner identifies medieval physiognomy as part of a broader move to understand both the logic of the medical symptom and the police operations of detection—a manifest inscription of a hidden, latent content: "Such a doctrine of signs, which claimed it was just a matter of looking at, describing, and interpreting signs correctly, coincided perfectly with the efforts that contemporary physicians were making to conceive and classify corporeal phenomena as visible *signa*; it also converged with the enhanced status conferred upon visible *warzeichen* and fact finding involved in the pursuit of justice and the development of techniques for establishing truth" (126). This analogy to medical practices attuned to reading symptoms can be misleading because it suggests that faces are simply present and available to sight, much like a swollen finger might indicate an underlying infection. Initially, we could recognize the added complexity that emerges when one acknowledges that a face can be disguised or at least be subject to calculation rather than arrive as a pure and unmediated appearance. More, the individual reading the face might possess a facility in reading signs somewhat less refined than that of a doctor of medicine. As Pearl notes in her consideration of a practice of physiognomy that remained a vital part of English social life well into the Victorian period, face reading operated as a "diagnostic technology" that mediated social encounters by establishing forms of informal and formal inspection (216). Pearl identifies physiognomy as "a surveillance mechanism" (27) and suggests that it was a social mode of looking at others "dependent on the experiences of both the observer and the observed" (26). As Elizabeth Craven knew a century earlier when she dreamed of the liberties a woman might acquire if she could travel in public unidentified, to speak of scenes of observation is to begin to capture a complex set of power relations that become attached to strategies of surveillance.

Physiognomy is not just a dynamic of looking at faces and regulating who goes where; it is a mode of regulating how individuals appear to one another that takes the form of looking at faces. Looking at others is never subjective, I would suggest. It is a practice guided by ideals and assumptions about what can be revealed by someone's physical appearance and presence in the world. Physiognomy insists on training the eye to notice particular details, recognize some matters and not others, and read them as representative of the person. As a technology of recognition, physiognomy invited individuals to judge whether someone was morally and socially desirable, an inveterate criminal, or a lesser human, and it proposed interacting with them on that basis. More than a technology of interaction, physiognomy was structured according to assumptions about classes, races, and ethnicities, and it cultivated the idea that what mattered most about another would be plainly visible. It became a "condition under which the visible is visible" and on that basis offers a glimpse of the ways in which social practices shape what one can recognize to be true (Galperin 8). For physiognomy, physical features became culturally coded and affirmed as an expression of one's essence. The body became the ground of the person and trained individuals to look away from other details as much as it trained them to focus on the lines of a face. Biometrics continues this tradition, and even if its technology is more sophisticated, its assumptions may not be.

The turn of the nineteenth century was an increasingly visual age defined by cultural spectacles, a newly responsive news media that made the public aware of events in the world they might never see, and by a new cosmopolitanism that filled urban centers with familiar- and unfamiliar-looking strangers. In this context, the desire to see and the capacity to make sense of anonymous others was new, vexing, and necessary for an emergent visual culture. This was an age in which, as Simpson has so powerfully documented, the poet William Wordsworth made a career out of two contradictory, yet synchronous, operations: he wrote in the language of the common person while he made a spectacle of the disappearance of common ways of life under industrial capitalism. Wordsworth wrestled with a matter that continues to undo us: the existence of a vibrant "democratic imperative (whereby we are all, or should be, equal) awkwardly conjoined with persistent or increasing economic and other disparities which cry out for address . . . but [which] seem impossible to redress" (*Wordsworth* 27). Physiognomy found a comfortable niche in such conditions, affirming the presence of recognizable others in ways that spoke to the era's expansions of democratic freedoms and desire to recognize more people as humans with rights and dignity than ever before. And it promised a perhaps too easy mode of managing one's social concern for those diverse others by making individuals legible as discrete entities. One could look at another and see the suffering written thereupon. Yet the very prospect of acknowledging, in some cases for the first time, the humanity of so many by looking them in the eye and attempting to gauge their character was a regulatory practice that tended to limit what one could see and understand to be the basis of shared human existence.

By attending to faces, physiognomy perceived people in ways that isolated and separated them, both one from another and also from an economic world that was so powerfully remaking their lives. Suffering could be individualized. Focusing on the face became a means of forcing so much relevant context to disappear from view and consideration. Such context is important because it vitally conditions what identity means, how individuals exist together, depend on each other, and are shaped by an unchosen social world of norms and ideals. Physiognomy prepared the way for assertions of *personal responsibility*, for example, that still seek to hold individuals to account for the social and economic forces that have impoverished them and which see their lives as largely disposable. In the present, such claims of individual responsibility loudly proclaim accountability in the name of equality often without readily attending to equality in circumstances, opportunity, means, privilege, or social vulnerability.

Physiognomy supported an emergent project of social self-definition in which the Romantic period's ascendant middle classes, including Wordsworth, increasingly established their own cultural force by identifying so many others as strange and as strangers compared to their own self-assertions of normalcy. Looking at strangers is always an operation of looking for strangers, a means of finding those others who were not so different as to be unrecognizable but whose strangeness would likewise help to define and consolidate the proper norms and ideals of a newly dominant class. For anxious middle-class citizens at the end of the eighteenth century, the promise to discern and describe the moral failings of individuals unlike themselves provided a powerful means of establishing a coherent class identity. Physiognomy directly supported norms that aggressively identified and policed the bourgeois values, behaviors, roles, and practices that would come to structure everyday life: such norms were established and perpetuated in conduct manuals and treatises on education, works concerned with natural science and moral medicine, and fiction, poetry, and drama. For a public concerned with degeneration, urbanization, and the policing of social difference, face reading offered a commonsensical and intuitive approach to discerning "the correspondence between the external and the internal man" (Lavater 1: 19) by interpreting the many elements of the face and held that the "beauty and deformity of the countenance is in a just, and determinate, proportion to the moral beauty of the man" (1: 183).

A quick examination of Lavater's *Essays on Physiognomy* shows how insistently appearance was regulated rather than recorded, manufactured rather than seen. Consider what Lavater has to say about the faces of women. Their faces are "highly sensible of purity, beauty, and symmetry," yet "the female thinks not profoundly; profound thought is the power of man" (3: 206). Outlining a technology of sight, Lavater is wholly unable to see women except to see them as less than men: "Man is serious—woman is gay. Man is the tallest and broadest—woman the smallest and weakest. Man is rough and hard—woman is smooth and soft. Man is brown—woman is fair. Man is wrinkly—woman is not" (3: 210). "Woman is not a foundation on which to build. She is the gold, silver, precious stones, wood, hay, stubble; the materials for building on the male foundation" (3: 209). If women have faces, I am quite sure Lavater cannot see them, given his nearly complete failure to exercise his capacity to look seriously and rationally—as only a man can—at women's faces.

Lavater was not unique in developing discriminatory norms that sought to condition what could be seen via acts of physiognomic inspection. In 1806, Charles Bell published *The Anatomy and Philosophy of Expression as Connected with the Fine Arts*. Bell used physiognomy to police the European against forms of nonwhite degeneracy. In a comment that anticipates phrenology—the racist sister-science to physiognomy—Bell contends "a perfect brain and a perfect skull are formed together" (144–145). The shape and angle of the face betray "national particularities" that Bell assembles into a proto-evolutionary scale, with the proper vertical profile of the European representing the norm. Deviating from this ideal norm of a vertically arranged face meant diminishing "the beauty and perfection of the form. For example, if the line formed an angle of seventy, it became the head of a Negro; if declining backwards still farther, by the depression of the brain-case, say to sixty, it declared the face of an orang-outang; and so, down to the dog" (27). While popular pseudoscientific understandings of the body show how knowledge is organized and structured within social practices—including practices of white supremacy—and thus reflect the relations of power in a given context, they are also part of a particular Enlightenment project that selectively humanized some human beings more than others. Indeed, physiognomy is a domestic expression of an increasingly common colonial anxiety regarding the policing of social differences evident in European cultural explorations of a diverse humanity. As literary scholar Felicity Nussbaum notes, the historical production of the human implies parsing "physical differences that Europeans 'discover' around the world" during this era, a process that inscribes a geography of humanity that fantasizes separation and difference amid a rapidly shrinking world of colonial empires (254).

Physiognomy bestows human life upon a face by making it an object that can betray, confirm, confess, and otherwise alienate the person it stands in for. The rhetorical term for this is *prosopopoeia*, which means, according to the leading figure of American deconstruction, Paul de Man, "to confer a mask or face" (76). A mask or a face? I am not sure that physiognomy can comprehend the difference, given the way that it seeks to constrain recognition and insists on seeing what is not there. If physiognomy promises face-to-face honesty that reveals one's character, perhaps in much the same way advanced biometrics promises a transparent relationship between the presence of a person and an identity, it goes to great lengths to both create what it beholds and to not see whatever fails to fit within its frames of recognition. Biometric technology can seem ordained by that master physiognomist, Perseus, who severed the head of the Medusa precisely by never looking directly at it. Isn't this what biometric thought always does? It sees without looking and severs the person from so much that matters for identity and existence and then mounts its biometric truth upon a shield in order to defend against other ways of knowing reality.

Would biometric thought have emerged if physiognomy did not exist? I am not sure. Clearly physiognomy is not the only source of biometric ideas of the body as a ground for identity, but it may be the source that best captures the optimism and promise that remain at the core of biometrics. Much of the cultural force of physiognomy comes from the way it relied upon a commonsense mode of social perception and an attractive logic of making sense out of unknown faces. The popularity and prevalence of face reading cemented its practices and made identification a matter of carefully constrained acts of social recognition—even while representing it as a matter of discovering what lay right there, visible to the naked eye.

To make individuals subject to the question of "Who are you?", as physiognomy did, was to place this fundamental query at the core of social life. But enforcing "the obligation to account for oneself in the sphere of public life" tends to reveal that this query can lead to many more outcomes than identification (Simpson, *Wordsworth* 25). Wordsworth's poetic investigations into the rural poor remind us that "Who are you?" can also elicit a more capacious understanding of "How is it that you live and what is it you do?" as the poet puts it in "Resolution and Independence" (128). Accounting for oneself and others may not lead to identity but could instead ask after the social conditions of existence. The response Wordsworth receives to his question, from even such an expanded consideration of the life of another, is revealing. It is worthwhile to read Simpson's powerful examination of the stakes of this encounter:

> This question, put by the poet-speaker to the old leech-gatherer (for the second time, since the poet wasn't listening to the answer the first time), is of the essence of modern democratic interaction, only slightly less aggressive than the famous Althusserian moment of interpellation, "hey you!" It says: identify yourself, explain yourself, account for yourself. The old man, again famously, is a model of civility, with his "courteous speech" and "demeanour kind." But the poet-speaker is tuned out, tripping, carried along by self-absorbed projections of the tragic lives of poets and the failures of the faculty of the imagination. There is no celebration of dialogue or conversation and no mutual accommodation, and it is to Wordsworth's credit that he does not pretend that there is. This is not the standard picaresque or charitable interaction in which some social bond is established between strangers that models or presages the initiation of a social contract. Instead, the poem presents an anatomy of how hard it can be to take that step. (*Wordsworth* 25)

Part of the appeal of physiognomy was its promise to avoid so much of the complexity that comes with actual human interactions of the sort Wordsworth contemplates. As Simpson notes, interactions that entail accounting for oneself may enflame rather than relax social tensions, and the poet finds little that feels good here and even less that bonds him to another. Where Wordsworth discovers the absence of a shared human nature and the inability to connect meaningfully with another, physiognomy reveals instead that we can and will connect with one another and understand them as we do ourselves. Yet that promise depends upon removing so much of the context that hangs here between Wordsworth and the leech-gatherer; it depends on acts of imagination, creating individuals within particular recognizable terms rather than apprehending them and their circumstances as they are. So, physiognomy not only produces individuals as knowable and insists that identity must always exist as something recognizable; it also makes Wordsworth's encounter unthinkable. Physiognomy cannot comprehend that an individual might not be knowable. Face reading insists that the difference and strangeness of the stranger will always be overcome.

It is fair to ask why I want to insist on the possibility of perceiving or reestablishing disconnection and alienation from one another when such features of social life might be among the least desirable. I do not mean to advocate for disconnection any more than Wordsworth did. I only want to acknowledge that biometric thought encourages us to recognize a scene of calculation, which loses so much of what makes us who we are, as if it were an expression of connection and human universality and the basis for social bonds. What we have been calling connection and respect is actually a means of annulling the difference of another by allowing an individual to appear only in terms that are readily comprehensible and fit within existing social schemas of identity. But perhaps physiognomy can make this problem heard. Perhaps it contains within itself the potential to illustrate or make legible the very operations that it so often seeks to make unrecognizable.

In the pages that follow, I consider how these tensions at the heart of physiognomy might be reflected in works of culture, works that shaped and reflected the social recognitions that dominated their eras and structured how individuals thought and acted in the world.

Consider William Blake's famous poem "London" (1794), in which he encounters his fellow citizens without ever identifying them or demanding they be subject to biometric imperatives:

I wander thro' each charter'd street,
Near where the charter'd Thames does flow.
And mark in every face I meet
Marks of weakness, marks of woe.
In every cry of every Man,

In every Infants cry of fear,
In every voice: in every ban,
The mind-forg'd manacles I hear
How the Chimney-sweepers cry
Every blackning Church appalls,
And the hapless Soldiers sigh
Runs in blood down Palace walls
But most thro' midnight streets I hear
How the youthful Harlots curse
Blasts the new-born Infants tear
And blights with plagues the Marriage hearse. (Blake)

Blake's poem moves from the visual to the auditory as he emphasizes the sounds that punctuate the streets of the capital at midnight. Readers are meant to register some compassion and offense at these rough sounds: they are sounds that should not be heard by children, nor should they be coming from the voices of children working the streets late at night. Blake does not wander through London in order to judge the morality of the lives he apprehends or discover the truth of those he meets. Physiognomy was so often prepared to do just that, but his regard for faces produces something distinct from those operations. We might imagine that the alternative is a sympathetic expression of social concern, lamenting the sorry state of the urban poor and their "marks of weakness, marks of woe." What Blake offers instead is a commentary upon the social process of making sense of those living in the public eye. It is the speaker of the poem who sees weakness in every face and hears in every voice defeating assumptions that have been forged within the mind. Blake dwells not on individuals but on the contexts in which they are recognized to be a plague, to be appalling, and to be blights upon the world. Such dehumanization belongs to the world the speaker inhabits, and it reflects the force of social understandings that regulate what it means to live in London, especially among those whose lives are so readily reduced to spectacles of urban despair.

I cannot help but hear exasperation in the poet's reception of "every cry of every man" and "every infants cry of fear" and "in every voice: in every ban." It is an exasperation, perhaps, with the power of pervasive social recognitions that tirelessly infiltrate and structure our psyches. The speaker's refrain identifies a homogenizing biometric impulse that acts as a mode of identification, as if every face revealed the same reality, and a means of not seeing the real and distinct lives present on the streets of London. For Blake, it was biometric thought itself that needed to be understood and seen at least as much as the lives it made legible. Physiognomic inspections of "every face I meet" were producing modes of identification that, while sometimes sympathetic, did little to comprehend the conditions that made individuals and communities legible, as if they were plagues. It is so telling, then, that Blake acknowledges but never sees or reads these faces, preferring only to recognize their humanity.

Blake had illustrated part of an English translation of Lavater's work, and this poem suggests he thought carefully about the world that these ideas were already producing. If it was difficult for Blake to witness the poor and economically abandoned and to find a way of seeing them that would not reproduce physiognomy's tendency to transform looking into dehumanizing identification, consider how much harder life was for one who directly bore the scars of such dehumanizing modes of thought, as was the case for an African living as free as was possible for an emancipated slave at a time when too many saw such a face as a walking contradiction in terms.

Olaudah Equiano has long been known as the author of the first slave narrative and a key figure in the international movement to abolish slavery that successfully convinced the British government to end the slave trade in 1807. *The Interesting Narrative of Olaudah Equiano* (1789) recounts his experience of growing up as a member of the Igbo tribe in modern-day Nigeria, being kidnapped and sold, the traumas of slavery that followed, and his remarkable survival. He eventually purchased his freedom and continued to work as a sailor on the ships that so powerfully defined the Atlantic slave trade and all the lives within its deadly compass. It is a reality that Equiano knew well, and to this day his narrative reveals a great deal about the resilience and spirit of an individual who refused to let others fail to see and recognize his humanity.

Equiano's account of slavery has recently been questioned, however, based on new evidence that suggests Equiano was born in South Carolina, and it is important to think about what this controversy reveals about the power of identification and the nature of its operations. Examining this controversy makes clear how powerful biometric thought remains, even within academic discourse, and begs several questions. How should Equiano be identified? What circumstances matter if one is to establish his identity? What can be seen and what is not seen when one inspects Equiano as if biometrically, by checking his identity papers?

In his remarkable biography of Equiano, Vincent Carretta makes a small note of two substantial pieces of evidence—a baptismal certificate and a crew log identifying the sailors aboard a British vessel—that prove Equiano was born in the United States, not Africa. A 1759 baptismal record in the name of Gustavus Vassa—which is the third name he received as a slave as he was bought and sold and named and renamed—identified him as a "Black born in Carolina 12 years old." Likewise, a 1773 ship muster for the *Race Horse* lists a "Gust. Weston" and a "Gust. Feston" of "S. Carolina." The evidence is slim but especially compelling for a biometric culture such as ours that trusts the power of official identification measures to represent rather than invent reality. But what do these facts reveal, and what do they make less comprehensible? What does this reveal about the operations of biometric thought two hundred years ago that remain relevant for too many today? And how did Equiano himself navigate a biometric culture that sought to recognize him as a commodity before it acknowledged him as a person?

The stakes of Carretta's discovery are significant for our understanding of history, but they might have had the force to alter history itself had these details been disclosed during Equiano's day. By providing a firsthand account, *The Interesting Narrative* promised to correct public narratives of slavery that misrepresented it as a gentle institution and sought to justify its continuation by negating any public memory of the Africans already suffering under slavery. Had Equiano lied about his experiences, it is difficult to imagine just how much this might have damaged the abolitionist cause. Truth was a battleground in these debates, given the sanitized descriptions many Britons received from pro-slavery writers addressing realities they would never witness. As Equiano once wrote to writer James Tobin in a letter in *The Public Advertiser*, a mild description of slavery depends upon not looking very closely: "You never saw the infliction of a severe punishment, implying thereby that there is none?" (331). The public debate around slavery was a matter of telling the truth and regarding what could be recognized as the truth. And this raises another interesting consideration.

Equiano mentions his baptism directly in the text; this is an unlikely gesture for someone with something to hide, especially when proslavery advocates would have seized upon any chance to show that this eyewitness was a fraud (Davidson 27). The claim that he was born in the United States transforms his book. It makes his account of the middle passage into a work of fiction rather than testimony and evidence. That too is a curious distinction. Given the terror and trauma that constituted the middle passage of the Atlantic slave trade, what does it mean to be a true witness? Is a true witness someone who recalls directly what happened, or should the truth be sought in testimony that reveals the incoherencies and inconsistencies that are the hallmark of trauma? Can the very form of telling—a form that might be shattered, imperfectly recalled, expressive of traumatic gaps and silences—reveal the truth of a person's experiences, or is it the factual basis of the account that makes it stand as a testimony of truth?

Does the evidence Carretta uncovered speak for itself? Any number of situations might explain the documents he discovered. If Equiano is indeed the person recorded as Weston or Feston, is it possible that his place of birth was also misheard, as Cathy Davidson suggests? For example, my mother-in-law's family name was Shaw before emigrating to the United States, where the German script on their documents was rendered Thau, and that has been their name ever since. Perhaps the reason is less arbitrary. Davidson wonders if Equiano chose South Carolina because being marked as African might have made him a greater target for exploitation and mistreatment. Much of his account of living as a freed slave suggests that the vicious force of racism made him especially vulnerable to mistreatment by individuals who had no interest in maintaining his freedom. If this is the case, it suggests that identity is something that may not be attached to a person and a body but something that can be wielded strategically and repurposed to fit the occasion in order to protect oneself. This might be especially true for a culture unwilling to recognize the legitimacy of Equiano's existence as a free individual and a human being. Better to live a well-chosen lie that makes one's humanity recognizable, perhaps, than risk the violence that comes to those who are illegible.

Does the baptismal certificate prove anything? Only that his birthplace is recorded as South Carolina. This is the only fact here. There is no way to know the truth of that assertion, why it was asserted, or even by whom it was given: twelve-year-old Equiano or his owners? Davidson spells out the significance of this empirical record by asking: What else does it say? Davidson reviewed the original baptism record and noted its difference from those that come before and after it:

Diana Stacey D. of William by Frances
Gustavus Vassa a Black born in Carolina 12 years old
Elizabeth Husbands D. of Wm by Mary 5 years old. (28)

Equiano has no mother and no father. This is "documentary evidence of erasure," not a document that verifies identity (29). At best, this "evidence illustrated . . . the closed loop of institutional" memory to assert as true the information it has itself created according to its own peculiar logic (Youngquist 182). This is the same logic that allows Equiano, unlike the individuals whose baptisms are recorded before and after him, to have no parents. This cannot be true, yet this is literally the truth to which the record attests.

Carretta's empiricism, moreover, cannot address Equiano's aspiration that his text will be more than "an eyewitness account of one slave's journey to personal freedom and spiritual salvation" (Youngquist 183). As a work with abolitionist aims in mind, Equiano's book has always been more than an account of himself. "The salient issues for me are epistemological and not simply factual," Davidson agrees. "What do we learn because of the specific questions we ask and, equally important, what do we not learn because we have chosen to ask those questions?" (18). If one sees recorded in *The Interesting Narrative* only a story of the author, then one has missed a great deal of what it reveals about empire, economics, religion, racism, power, and how individuals are surviving the black Atlantic.

What Equiano knows so well is that some people have been dispossessed of their identity, as a very feature of their being, and this truth cannot be captured by a process of punctual identification. Biometric thought may be entirely incapable of comprehending a life like Equiano's yet also strangely familiar with its dispossessions. *The Interesting Narrative* is not an account of overcoming the unchosen conditions that govern his identity but rather a mode of living despite being compelled to make himself legible within the terms of a society entirely unprepared to recognize his existence. If we expect official records to tell us who he is, we are allowing them to eliminate the reality of a life that was sometimes defined by the power of others to decide who he was and what he would be. What is at stake is nothing less than the decision to remember the full extent of the violence of slavery or to let that history disappear as irrelevant context because we mistakenly believe that individuals control the conditions that govern how and in what ways their experiences become recognizable.

For Equiano, biometric thought was a necessary component of his life as a slave who had purchased his freedom. He routinely asserted his documented existence, brandishing his manumission papers. Documented identity could not guarantee his freedom, but it might be effective in some instances to keep him free. Such documents came at a cost. They replaced a past of dispossession and loss with presence and facts, reflecting the power of British imperialism to define the terms within which he could be recognized to exist. Biometric thought does not erase that past from Equiano's memory nor from the public record. But by failing to acknowledge the terms within which it makes identity legible, biometric thought creates an antagonism in which readers may be invited to decide upon the truth of Equiano's account. His remarkable autobiography can, in an instant, be thrown into doubt by the discovery of some slender facts that carry the weight and force of official knowledge and the illusion of authority that comes as an effect of documentation. This remains the ruse of biometric thought: it never destroys other ways of knowing oneself or others; it never claims to be more than a discrete procedure used for particular strategic purposes. But its effects are profound, and its power to produce the very thing it appears to represent transforms what we can know and how we recognize and live with identity as a powerful feature of existence.

If much of this book has been concerned with biometric thought in the present and its effects on individuals whose lives are more or less structured by its operations, Equiano's narrative reminds us that these effects may outlive the individuals involved and make the nature of their existence illegible and incomprehensible well into the future. If biometric thought can sometimes seek to police not only the living but also what one can know of the dead, it can likewise be a force that prompts one to reconsider what can be seen in the face of another. Is the face of another necessarily something one can see? To what extent does biometric thought look at another in order to see not what is there but what one wishes to see there? These are some of the questions that structure Percy Shelley's remarkable poem "On the Medusa of Leonardo Da Vinci," a work that reminds me of the importance of thinking carefully about the social priorities and presumptions that almost always structure what it means to look at and identify another.

In the autumn of 1819, the poet Percy Shelley composed a series of works that sought to understand England at a time of social upheaval and economic depression. These poems bore witness to the social protest by sixty thousand on August 16, 1819, at St. Peter's Field in Manchester, which was violently suppressed by the state. The event came to be known as Peterloo, a name that evokes the shame felt by a nation that had once defeated Napoleon at the battle of Waterloo and which now turned its might against its own population.

Shelley married into what was perhaps the First Family of physiognomy in the Romantic era. His wife, Mary Shelley, was the author of *Frankenstein*, which we have already noted is a powerful rebuke to biometric thinking, insisting as it does on the centrality of social recognitions that regulate how one exists in the world. Her parents were the prominent public intellectuals Mary Wollstonecraft and William Godwin. As literary scholar Michelle Faubert notes, physiognomy attracted Wollstonecraft's attention but ultimately proved to be irreconcilable with her attachment to education as that which "forms, and can reform, human character," and this "explains, at least in part, why Wollstonecraft lost interest in her physiognomical projects" (294). Godwin's fiction relied on a visual code that regularly pondered whether facial features and physical appearance were indicators of character (Juengel 74). Godwin even had little Mary's face read by a trained physiognomist. Given the prominence of biometric thought within his immediate family and wider culture, Percy Shelley was primed to notice the destructive force of this way of thinking for the events in Manchester. Nor was he alone in perceiving how much identification matters when a nation seeks to perceive its own citizens as hostile enemies.

In the aftermath of Peterloo, Lord Castlereagh asserted in the House of Commons that the protestors were the cause of the violence brought against them by the state at St. Peter's Field: "The magistrates had not intended to interfere with the meeting. They had taken their post for the purpose of watching the meeting, not of breaking it up. After a variety of depositions had been made, which give a character of terror to the meeting in the minds of the people of Manchester, and which gave the meeting that illegal character which the law asserts, then had the magistrates granted a warrant" (Stewart 827). The citizens had assembled to protest laws that made grains and flour unaffordable at a time of mass unemployment and economic distress. Marginalized, they wanted the right to vote and the right to representatives who would take their basic need for survival into account. They faced a government unwilling to even ask, as Wordsworth did, "How is it that you live?" Yet because their protest could be recognized as terrorist—and we should remember that it is this era of European social upheaval that gives us the modern meaning of terrorism, thanks to the Terror in France under Robespierre in the years following the French Revolution in 1789—these individuals had no dignity, no rights, and represented lives that could be violently ended.

A cavalry charge on the assembled crowd killed eleven and wounded four hundred. Reading of this event while in Italy, Shelley was incensed: "The torrent of my indignation has not yet done boiling in my veins. I wait anxiously [to] hear how the Country will express its sense of this bloody murderous oppression of its destroyers" (Shelley, *Letters* 117). During this period, he penned "England in 1819," "Song to the Men of England," and "The Mask of Anarchy," in which he described the progress of tyranny across England and counsels the "terror-stricken" laboring classes (55) to "stand ye calm and resolute" in a "sea of death and mourning" and maintain faith in the laws of England (319, 318). Shelley hoped that citizens would not respond with docility but also that the nation would not devolve into sheer chaos.

During this same tumultuous autumn, he composed "On the Medusa of Leonardo Da Vinci." The poem was written after viewing a painting of the Medusa's severed head, a work of art believed to be by Da Vinci and held at the Uffizi Gallery in Florence. The poem is concerned with terror and death, though perhaps not in the way we might expect given the very different meanings associated with the Medusa in Shelley's day. The Medusa has long been a figure of terror associated with the horrors of anarchy and revolution in the English mind. Yet for many in the early nineteenth century, the Medusa was celebrated as a figure of political unrest. She was a woman unjustly disfigured, a perception that remembers it was Athena who transformed the young woman into a creature cursed to destroy every being she looked at as a punishment for having been raped by Poseidon in Athena's temple. In the radical labor press of the 1810s, publications such as *The Medusa* and *The Gorgon* claimed her as an abject heroine for England's disenfranchised classes, who had been abused and defaced by the prevailing structures of power and authority. As Jerome McGann notes, "For a poet inclined to interpret, in a radical way, certain traditional myths like the fall of the angels and the binding of Prometheus . . . Shelley would not have been able to see [the Medusa] as anything but a victim of the tyranny and cowardice of established power" (7).

So, what we have in the myth of Medusa is a narrative of despotic power and the ways in which it perversely recognizes its own violence as a justification for further destruction and abuse, the result of which is a creature who looks back from such abuse and imperils the lives of everyone she encounters. It was in this context that Shelley connected the triumphant victory of a nation over its citizens and Perseus's success in exterminating the Medusa:

It lieth, gazing on the midnight sky,
Upon the cloudy mountain peak supine;
Below, far lands are seen tremblingly;
Its horror and its beauty are divine.
Upon its lips and eyelids seems to lie
Loveliness like a shadow, from which shrine,
Fiery and lurid, struggling underneath,
The agonies of anguish and of death.

Yet it is less the horror than the grace
Which turns the gazer's spirit into stone;
Whereon the lineaments of that dead face
Are graven, till the characters be grown
Into itself, and thought no more can trace;
'Tis the melodious hue of beauty thrown
Athwart the darkness and the glare of pain,
Which humanize and harmonize the strain.

And from its head as from one body grow,
As [] grass out of a watery rock,
Hairs which are vipers, and they curl and flow
And their long tangles in each other lock,
And with unending involutions shew
Their mailed radiance, as it were to mock
The torture and the death within, and saw
The solid air with many a ragged jaw.

And from a stone beside, a poisonous eft
Peeps idly into those Gorgonian eyes;
Whilst in the air a ghastly bat, bereft
Of sense, has flitted with a mad surprise
Out of the cave this hideous light had cleft,
And he comes hastening like a moth that hies
After a taper; and the midnight sky

Flares, a light more dread than obscurity.
'Tis the tempestuous loveliness of terror;
For from the serpents gleams a brazen glare
Kindled by that inextricable error,
Which makes a thrilling vapour of the air
Become a [] and ever-shifting mirror
Of all the beauty and the terror there—
A woman's countenance, with serpent locks,
Gazing in death on heaven from those wet rocks.

Shelley included a final stanza that was discovered years later. Though the entire
work may be fragmentary—whether by design or not, the occasional word is
missing—it is difficult to imagine the poem closing without these final lines:

It is a woman's countenance divine
With everlasting beauty breathing there

Which from a stormy mountain's peak, supine
Gazes into the [] night's trembling air.
It is a trunkless head, and on its feature
Death has met life, but there is life in death,
The blood is frozen—but unconquered
Nature Seems struggling to the last—without a breath
The fragment of an uncreated creature.

Without delving deeply into the complex nuances of this poem, I want to note the persistent humanizing force of Shelley's consideration of "a woman's countenance divine." The poem is less an identification of who the Medusa really is than it is an effort to grasp the tentativeness and uncertainty felt by the speaker who remembers the force that biometric thought brings to bear upon those whose identities are created by such productive normalizing operations of inspection. Perseus has left, and the viewer remains to contemplate this scene of destruction wrought by physiognomy's power to behead an individual and yet never see who is truly there. The poem remembers the singular humanity that has been submerged by the Medusa's transformation. Shelley insists on remembering the context that has made her what she is and sees "the melodious hue of beauty thrown / Athwart the darkness and the glare of pain, / Which humanize and harmonize the strain." If biometric thought can seek to destroy others in the name of terror felt and pleas for security, this is not all its operations might apprehend, Shelley insists.

As was the case for Blake, Shelley finds that biometric thought can reveal the social construction of monsters and the thinking that, in turn, justifies the destruction of the very monsters a society has created. This was true at Peterloo just as it was for the Medusa. If we see in biometric thought only the guarantee of identity, we will continue to miss what Shelley reveals so powerfully: that biometric thought produces a social scene of identification that lays bare the ways in which identity depends upon relations of power rather than the presence of some natural inborn thing or quality. Is the Medusa a monster or a victim? What we see depends entirely on how she is framed and what we are invited to see there. What Shelley would like us to remember is that whatever we can see and behold thanks to the operations of biometric thought, there remains an "uncreated creature" beneath and before any of these social recognitions. Human culture and its operations of knowledge and power create the possibility of identifying another and the terms within which identification will be possible as well as what must be forgotten as a condition of that possibility.

If biometric thought has a history, this means that its present and its future can be shaped and shaped differently. As persuasive as physiognomy was two hundred years ago, its operations were contested and its effects were rarely absolute. Powerful and thoughtful voices such as the writers and artists I have mentioned here have long sought to understand identity as a richly woven social experience that binds us meaningfully to the lives of others while acknowledging that conceiving of identity as more than the physical presence of an individual instantiates rather than resolves any number of social tensions and complexities. When we follow biometric thought and accept identity as something attached to the discrete individual, we forget these rich traditions of thought and the ways in which individuals are presently living with and despite the conditions of biometric thought. Such lives are direct challenges to the claims that we identify, rather than annul, the existence of another by looking to what can be verified according to physical and biological features. This very brief history of techniques of identification shows that the present operations of biometric thought are far from the only ones imaginable. What would it mean to imagine modes of human recognition that do not generate punctual scenes of identification nor dispossess so many of their lived attachments to the world and culture? Is it possible to account for the ways in which we have become created creatures?

6 · THINKING IN THE WAKE OF BIOMETRIC THOUGHT

In 2014, the graffiti artist Banksy added a stenciled image to Clacton-on-Sea in Britain: a colorful African songbird juxtaposed against the ugly desires of inhospitable colorless birds carrying placards and banners that read "Migrants not welcome," "Go back to Africa," and "Keep off our worms." Banksy's artwork is known for drawing attention to how social matters are framed, and in this case the image encourages viewers to rethink how they understand migration. While the image did not last long before being covered up, it references a set of hostilities toward immigrants that has remained volatile in the United Kingdom and was key to the 2016 referendum decision to leave the European Union.

By taking on human activities and protesting migration, the birds quickly appear obscene as they try to prevent the arrival of a vibrant companion engaged in the essence of life for a migratory animal. It powerfully rethinks a biometric logic that can only perceive strangers to be unwelcome, unwanted, or dangerous. The image reminds viewers, moreover, that animal life does not care for political boundaries or the walls sometimes erected to mark them and highlights that some concerns with migration may refuse to acknowledge that we all exist on this planet together as earthlings.

What could be more natural than migration, the image seems to ask? Biometric thought may not imagine such a possibility or acknowledge the histories that support it, I suspect, and this powerful reconceptualization of migration targets the assumptions individuals hold about who naturally belongs in a given territory and who does not, especially for an English community that elected the first member of Parliament for the UK Independence Party shortly after this image challenged that party's anti-immigration policies. How far can an argument grounded in nature go, however? Human migration in present times is rarely natural. Today, civil war, sudden drought, instability, economic dispossession, flooding, persecution, and many different desires and fears lead individuals to migrate elsewhere in search of a livable life, temporarily or permanently. These are not the seasonal journeys of birds.

FIGURE 6.1. Photo of Banksy graffiti in Clacton-on-Sea, Britain. (Reproduced with permission of Universal News and Sport.)

Perhaps Banksy's image suggests that African migrants naturally want to find themselves in Europe. Is their destiny to be migratory, as if home were necessarily and permanently insupportable? Is Europe naturally desirable? The truth of the matter may actually be that migrants need to verse, much like Makina did in *Signs Preceding the End of the World*, and where one ends up matters less, so long as it better supports life. What if we read this image against its intention and consider the possibility that what we see here is a flock of protesting birds that have traveled abroad in order to protest the potential movement of others. The image does not exactly make sense in this context, but reading it like this highlights the image's assumption that some birds are naturally and permanently at home in one location, while other birds must migrate and are met with hostility when they do so. The logic of the piece depends on seeing migration as natural. Some birds migrate and others do not. From an ornithological perspective, this is true, but as a metaphor this means segmenting the world into those who can reside at home and those who are permanent migrants. In practice, human migration rarely comes without some degree of violence or neglect, and a metaphor of seasonal migration does not seem adequate to capture these relations of power.

Banksy's image raises a number of possibilities for thought, then, but I want to condense these into a single query that I will consider in this chapter: How do we think in the wake of biometric thought? This broad question is focused by attending to several related analytical possibilities that arise as alternatives to the organizing and regulating force of biometric thought. I do not examine these in order to determine which one provides the best or most realistic response, or which can be implemented most readily in hopes of ameliorating biometric thought's worst offenses, or which most definitively alters how we presently regulate access and human identification. Instead, I approach these as modes of not thinking biometrically. I proceed in this inconclusive manner in part because I have argued that there is no single source or single goal for biometric thought and thus possible points of revision and contestation are as varied as its points of attachment to the world. Indeed, there are good arguments for deployments of biometric thought that may support and enable people to live. In what follows, I want to signal briefly the value of thinking in ways that seek to be as responsible as possible to the lives and realities that biometric thought recognizes and fails to recognize in its operations.

Activists around the world are supporting migrants, helping to amplify their voices, and working to create conditions in which they will be heard. Lawyers are representing migrants whose survival depends on versing and now confront legal systems defined by documents and trails of paper, not worn shoes and incommunicable trauma. Individuals are routinely responding in formal and informal ways to the pressing needs felt by migrants arriving to new homes or versing through a country, sometimes by helping someone find a home and sometimes by leaving jugs of water in the desert. Their actions show how we retain our humanity in the face of a biometric mode of thought that can think only of regulating movement, not of human dignity or justice.

Biometric thought may make our shared humanity unthinkable as it encourages us to abandon others to their discrete, individual lives and to abandon the very idea of a shared planetary existence in which we all depend on one another and create the conditions necessary to flourish. Biometric thought often makes impossible conditions worse and even inescapable. And Banksy's image reminds us that new ways of thinking and understanding migration may produce new problems, too. But this is not a strong argument against such provocations. Instead it is a sign that thinking otherwise is a commitment to continue thinking.

What might it mean to think and live in ways not endorsed and constrained by biometric thought? Above all else, it means no longer recognizing individuals as if they are alone in the world. It means thinking about the social conditions within which one exists and which makes life possible, enjoyable, difficult, bearable, and unbearable. All of these social relations that biometric thought is positioned to ignore or recognize to be irrelevant must instead be recognized to matter. This does not make problems easier; in fact, I suspect this means problems become harder to solve. Biometric thought tends to ignore or suppress any number of relations of power, especially when it presumes that we are all equally in possession of ourselves and all equally affected by the demand to authenticate our identity.

Assessing and acting in situations structured by relations of power and compet-
ing claims and interests as well as differences in perspective and experience is no
simple matter. But ignoring such complexity, as biometric thought does, without
accounting for the effects of such reductions of reality to procedural fantasies of
punctual identification, is not an adequate answer to these challenges.

Biometric thought encourages one to understand that global economic forces
are unrelated to human movement and foreign policy activities are inconsequential for what goes on at borders. Yet surely if individuals can be made to be legible
in ways that reveal their legal identity, their perseverance and survival can also signify the traumas, harm, abandonment, persecution, and hope that have compelled them to migrate and seek access to another territory. We cannot continue
to believe that deprivation leaves no marks or could never be made recognizable.

Refusing biometric thought would mean understanding the world in terms
other than those of borders and separation, foreigners and strangers. Borders
increasingly only apply to people, not the social and economic abandonment that
wrecks lives in the name of austerity, obscene profits, or transferring public goods
into private hands. If the rallying cry of *No nations, no borders* sounds impossibly
utopian, one would do well to recognize that this is the corporate utopia being
created by the leading edge of capitalism and neoliberalism. A practical model for
border-free existence is already in place and is defining how many already live,
whether they like it or not. Under what notion of justice can we imagine that we
will directly support such movement only for the benefit of finance capital, corporate profits, and strategies of tax avoidance?

Borders help to create illusions of separation belied by economic realities that
intertwine the lives of those who make goods for export and those who consume
them. Refusing biometric thought means braiding together those who breathe the
toxic by-products of manufacturing with those who can pretend their consumer
goods never had to be manufactured. Global differences separate individuals in
very real ways in terms of protections for human rights, labor laws, environmental laws, health care, and so on, but biometric thought encourages us to see such
differences as absolute rather than acknowledge that they serve a contingent political function in a world where the life of one person depends materially on how
someone else lives in another part of the world. We cannot disavow our unchosen economic bonds with one another and our responsibilities to a finite planet,
however much biometric thought wants to reassure us that the foreigner still exists.
Can we say that anyone is truly foreign anymore?

This does not mean we ought to lose the differences between people and cultures and all become the same. Quite the opposite: it means insisting on differences and thinking about how those intersect with relations of power to make
some differences matter more than others and how they can be more consequential for some than for others. Justice cannot be grounded in a mission to normalize all experiences as essentially the same. This would only replicate a biometric

logic that pays little heed to the way differences come to matter and works with an idea of detectable individual physical differences that disavows differences in circumstance and experience.

Some will still hear that my argument—critical of a logic of biometric reduction that transforms complex social problems into individual matters of identity verification—amounts to an argument for lawlessness and a loss of control. Lawlessness is already the reality of many migrants who are driven further underground and toward more dangerous paths of exit and entry because they seek, for any number of reasons, to avoid biometric inspection. What is desperately needed is humane attention to how biometric thought is making lives less secure, especially when its operations target the most vulnerable among us, whether by criminalizing welfare recipients as potential cheats, isolating migrants as the reason for any number of social ills, or refusing to acknowledge the central importance of culture and history in defining how one lives.

I began this chapter with Banksy's image and a call to think in ways not contemplated by biometric thought. I proceed in this manner because I am keenly aware that laying bare the ugly indignities of biometric thought may not end expressions of violence but may instead create new barriers to movement or new experiences of vulnerability. My hope is that this is not the case, but it is an unsecured hope. Clearly those detained at Nauru who chose suicide in the wake of the Australian Supreme Court's 2016 recognition of the illegal status of that country's offshore immigration-detention centers did not experience an end to their feelings of hopelessness despite having their dire circumstances recognized.

Not every act of critical assessment of biometric thought will be equally effective in addressing the ways in which it produces and regulates an idea of human identity. Indeed, many critiques of biometrics produce understandings that enable it to continue largely without interruption because they reproduce many of the key assumptions that structure biometric thought. I am thinking, in particular, of those critiques that generate narratives of dystopian surveillance by associating new technology with the destruction of an older authentic way of life, a lament similar to the one Wordsworth offered many years ago. Consider Giorgio Agamben's comments, which I include at length and which exemplify some of the more thoughtful versions of this dispiriting technological sublime in which biometrics upends life as we know it:

> Anthropometric techniques that had been designed for criminals remained their exclusive privilege for some time. Even in 1943 the US Congress rejected the Citizens Identification Act, which aimed at instituting mandatory identification cards with fingerprints for all citizens. Nevertheless, by the rule that stipulates that what was identified for criminals, foreigners, or Jews will sooner or later be invariably applied to all human beings as such, techniques that had been developed for recidivist criminals began to extend in the course of the twentieth century to all

citizens. The mug shot, accompanied at times by fingerprints, became such an integral part of the identity card (a kind of condensed Bertillon card) that it gradually became obligatory in every state in the world.

But the extreme step has been taken only in our day and is still in the process of its full realization. Thanks to the development of biometric technologies that can rapidly obtain fingerprints and retinal or iris patterns by means of optical scanners, biometric apparatuses tend to move beyond the police stations and immigration offices to penetrate the sphere of everyday life. The entrance to the high school cafeteria, even in elementary schools in some countries (the industries of the biometric sector, which are undergoing a frenetic development, recommend that citizens get used to this sort of control from their early youth) is already regulated by an optical biometric apparatus, on which students distractedly place their hands. In France and other European countries, a new biometric identity card (INES) is in the making, which has an electronic microchip containing basic elements of identification (fingerprints and digital photos), as well as a signature sample to facilitate commercial transactions. As part of the unstoppable drifting of political power toward governmentality—in which a liberal paradigm curiously converges with a statist paradigm—Western democracies are preparing to establish an archive containing the DNA of every citizen, as much to ensure security and repression of crime as to manage public health.

Our attention is called from various quarters to the dangers embedded in the absolute and limitless control of a power that has at its disposal the biometric and genetic information of all of its citizens. With such power at hand, the extermination of the Jews (and every other imaginable genocide)—which was undertaken on the basis of incomparably less efficient documentation—would have been total and incredibly swift.

Even more serious, inasmuch as it has been completely unobserved, are the consequences that the biometric and biological identification have on the constitution of the subject. What kind of identity can one construct on the basis of data that is merely biological? Certainly not a personal identity, which used to be linked to the recognition by other members of the social group and, at the same time, on the capacity of the individual to take on the social mask without, however, being reduced to it. If, in the final analysis, my identity is now determined by biological facts—that in no way depend on my will, and over which I have no control—then the construction of something like a personal ethics becomes problematic. What relationship can I establish with my fingerprints or my genetic code? How can I take on, and also take distance from, such facts? The new identity is an identity without the person, as it were, in which the space of ethics as we used to think of it loses its sense and must be thought through again from the ground up. Until this happens, it makes sense to expect a general collapse of the personal ethical principles that have governed Western ethics for centuries. (50–52)

What kind of ethics is possible if identity has been replaced by biology and biology converted into data? What kind of future does this imagine? Recall the uneven and inconsistent history of identification over the past two hundred years, and remember that it is not a history of ever-increasing forces of surveillance and control. Another path forward may yet be possible, and there can scarcely be more evidence it is necessary. The question of ethics that Agamben raises in this context is one that I am keen to think through because ethics is rooted in one's relation to others and to the social world that forms individuals and creates the conditions and categories that guide ethical thought. Both of these overlapping domains are affected by biometric thought, though perhaps not in the way Agamben suggests. What kind of ethics is presupposed by biometrics, and what kind of ethics does it make more or less possible, perhaps despite its most public desires to guard against strangers? This is a slightly different question from the one Agamben poses; in order to pursue this thought, I need to shift away from the manner in which Agamben frames biometrics.

Agamben is interested in seeing an analysis of biometric thought arrive at a threshold of what he calls "the reduction of man to a naked life" (52), a phrase that evokes a reduction of oneself to biological life. The implication appears to be that this mode of life is pre-social, such that one is no longer possessed of a personal identity premised on "recognition by other members of the social group." He wonders what kind of ethics is possible if the specificity of the individual person has been evacuated and replaced by genetic code. For Agamben, ethics involves a social experience of individuality; it is premised not on a fantasy of absolute autonomy and control over the conditions of one's identity but on social conditions that govern who can be recognized and what they can be recognized to be. He identifies this as the "capacity of the individual to take on the social mask without, however, being reduced to it." The problem with biometrics, then, is that it replaces a social existence with a biological one.

What is the nature of this social existence? Is it defined by what Butler calls "the unchosen conditions" of one's life (Giving 19), conditions that structure and make possible the emergence of a functional and social concept of oneself? It is not clear how far Agamben is willing to go in acknowledging that we are dispossessed by the social conditions that make identity possible, because while he acknowledges this context, he also implies that this individual is willfully in control of these social masks. Biometrics decisively alters the individual's relation to a social world, he suggests, by remaking the body into something entirely beyond one's control and design. An individual no longer chooses one's appearance under biometric regimes.

I am wary of this before-and-after logic here, as I have been throughout, because it helps to erase the ways in which biometric thought depends upon more than new technologies. Biometrics is a mode of truth made possible by its attachments to strategies and tactics of veridiction or modes of ordering and recognizing reality that are selective and designed to forget what they do not acknowledge. I am also wary of the inference that, under a pre-biometric regime, one simply chooses a social mask. For Agamben, willfully selecting a given mask may not acknowledge the necessity of navigating social codes of legibility and appearance. If this is the case, then what he offers is a fantasy of choice in contrast to an unchosen compulsory biometric identity. There may be very different police powers attached to biometric legibility from those attached to everyday social codes of appearance, but I am not convinced they diverge from one another in an absolute way. By virtue of one's race or gender, for example, one may be masked in advance, as it were, by a society that infers knowledge on the basis of particular identity categories. Biometrics, moreover, could not function without existing social modes of identification that frame how we make sense of one another and oneself.

Identity always depends upon social conditions of emergence that regulate how one is recognized as an individual in the first place. If these conditions are biological or they involve navigating particular modes of normative public appearance, identity remains dependent upon a social narrative of existence. Even a concept of biological life has a social existence within a culture and constitutes a particular way of ordering and recognizing which details associated with one's existence will be identified as biological. This does not mean that all life on the planet is reducible to human interpretation and culture. Life will always have an existence outside of our modes of recognition. But Agamben naturalizes the human body in a manner that seems to ignore the extent to which it, especially, is subject to human regulation and human practices of knowledge regardless of whether it is conceived of as a biological or social entity.

His argument is that individuals lose control over their identities under biometric modes of perception. I am not convinced control was ever possible. More, what occurs under biometric thought is not a strictly biological existence but a social experience of identity that treats biological identity as a recognizable and mediating force by which to represent the core of an individual. In no way do biometric practices bypass culture and arrive at unmediated biological life, I have argued.

My most central concern, however, is this: Agamben's criticism follows biometric thought's gaze and settles upon biological existence as its decisive invention, robbing the individual of the ability to be oneself. He calls this "identity without the person," suggesting that biometric identification replaces the person within the physical body. Longing for the individual is less an indictment of biometric thought than a confirmation of its operations, however. Seeing nothing but individuals is what biometric thought achieves so effectively and so dangerously. When it focuses upon the body, it does so not to lose the person but to lose everything that makes a person possible and place the individual as if against a blank backdrop, and this actually includes losing a rigorous understanding of biological existence. There is no life as such without the conditions that make life possible. This means acknowledging the chosen and unchosen social relations that define existence: the land upon which one lives, the organisms that live with and within one, the air and food and water one requires, and the living relations to human and nonhuman life—all of which testify to the impossibility of existence flourishing in isolation. Biometrics always loses these realities of life and the interdependencies they entail in favor of brutal individuality.

Inventing a new form of body-oriented identity might well be promising, but such awe conspires with biometric thought to focus on the individual and fails to notice that it might continue to annihilate the world and all that it brings to bear upon how we live, including why someone crosses a border, seeks refuge, requires a connection to the land, feels attached or disconnected to others, and so on. Rather than something new, I cannot help but think that we already know exactly what ethics looks like under biometric thought. It is what we routinely hear issuing from every call for personal responsibility; every call to ignore the world and its role in shaping our lives; every call to look out for oneself; every call to make others responsible for the racism, sexism, ableism, or homophobia directed at them. Butler identifies this "entrepreneurial ethic" as part of a "war on the idea of interdependency" (Notes 67), which aspires to annul the social world that sustains us. A biometric ethics that sees a world without social relations is precisely part of this assault, in which individuals are made to claim, as their own, conditions they never chose but are asked to accept and affirm as an expression of who and what they are. It is not a person without content, as Agamben suggests. It is a person who is the only content in the world.

Thinking in ways no longer allied with biometric thought may prove challenging and may amount to nothing less than "the prospect of living by other laws," as Simpson puts it in another context (Romanticism 247). What might it look like to try to awaken from the dream of biometric thought, and what new social allegiances and alliances might become more possible as a result? The path forward begins with the question of ethics that Agamben has already raised.

I am not convinced that a worthwhile ethics can be rooted in procedures of identification and acts that seek to overcome the strangeness of another by reducing his or her difference to punctual determinations of identity. As strange as it might sound, what are needed are ethical obligations that depend upon the recognition of another's strangeness, not familiarity. On such a basis, one does not demand that another conform to my desires or to my perceptions. It does not look for humanizing representations of legitimate, worthy migrants. It does not seek to see familiar faces. Instead, it suggests that another has dignity and an existence that merits respect regardless of what I might know or be able to determine about this being. Such an ethics involves refusing to isolate individuals from social contexts and instead recognizing the schemas of intelligibility that mark some as illegal and others as familiar. It is not an effort to produce better, more equitable knowledge about who someone is, though this might matter, too. It is an ethics that has much in common with Levinas's claim that ethics entails a "disinterestedness" that he designates "the encounter with the other person's face" (229). While this can immediately sound like an expression of biometric recognition—and no doubt part of its rhetorical force leverages that history of thought—what Levinas has in mind is precisely not biometric operations of substitution in which the other is transformed into something identified and known. Instead, he is interested in the responsibility that begins and is sustained by experiences that do not belong "to a thematization of knowledge" but which instead take the form of a recognition of "the uniqueness of an irreducible I" (233). To see another as irreducibly singular is to see one as different and to stop exactly there, refusing further biometric translations that would seek to know them according to some mode of intelligibility. There may be occasions when biometric technology does exactly that. But biometric thought almost never stops there, which means that an instance of biometric technology is often part of a larger social schema, for example, that tracks welfare claimants using fingerprints as if they were criminals (Magnet 72) or treat migrants as threats to security.

For Levinas, seeing the face of another in such a presumptive manner is always an attempt to annul difference and destroy it by dehumanizing another's singularity. As a thinker committed to an ethics of difference, Levinas challenges a priority on the self and insists that seeing another produces a state not of knowing recognition but a recognition of "being outside-of-oneself before the face" of another (233). Butler's thought stands as an urgent reflector of this initial beacon by recognizing the implications in Levinas's thought and extending it to consider that the difference of another is also an opportunity to acknowledge conditions of existence that are beyond the self. To be a self is to exist with others and with responsibilities to social existence as such, precisely because there is no self without the experience of being outside of oneself that arrives in the presence and absence of another. The self is always a social category, moreover, that involves unchosen social forces that not only define us but imbricate us in a world of others. Identity, she notes, means thinking about "the permeable border" between a self and another, a thought that insists the self is always formed by what is beyond its bodily borders, and this means thinking about more than just a rhetoric of security lost to "invasion, encroachment, and impingement" but also forms of coexistence that appear as a result of being equally "vulnerable to destruction by the other, and in need of protection" (*Frames* 43). Whether one wishes to acknowledge it or not, "we are bound to one another in this power and this precariousness" (43). The thought of the borders of oneself, then, is a thought that must consider what sustains a being and how this depends on "the permeability of the border" (43). Such a metaphor sounds strange given the ways that our flesh acts as a border. But then one can remember that a body plays host to other organisms, can create life, fracture a sense of self, or make it all the more certain. Add to this all of the conditions outside of oneself that one requires to flourish and even survive, and it becomes clear that "my existence is not mine alone, but is to be found outside myself, in this set of relations that precede and exceed the boundaries of who I am. If I have a boundary at all, or if a boundary can be said to belong to me, it is only because I have become separated from others, and it is only on condition of this separation that I can relate to them at all" (*Frames* 44). Embodied existence of the sort that biometric thought insists upon is not nearly as discrete or as individual as it appears. To be a body depends on any number of "institutional structures and broader social worlds. We cannot talk about a body without knowing what supports that body, and what its relation to that support—or lack of support—might be. In this way, the body is less an entity than a living set of relations; the body cannot be fully disassociated from the infrastructural and environmental conditions of its living and acting" (*Notes* 64–65).

Attending to the conditions that govern how individuals appear has been at the core of this book, and the alternative path that I am proposing here is a mode of becoming attentive to what exceeds those conditions and what is not represented or recognizable thanks to those conditions and the relations of power that support them. It means being responsive to the very real possibility that what biometric thought knows about the world is not all there is to know and that it leaves out a great deal more than it captures.

What might this look like? How can we imagine an ethics that depends upon difference when security and sovereignty have become manifest in acts of biometric identification? Thinking about ethics in the wake of biometric thought compels us to consider ways of apprehending the world and one another that are not reducible to operations of punctual identification, access control, and permission. While biometric thought presumes to make only certain matters recognizable, it nonetheless cannot prevent unsanctioned recognitions of the sort that may lead toward other determinations of identity, other determinations of the social basis of existence.

In 1795, Immanuel Kant published "Toward Perpetual Peace," an essay that powerfully reflects a time when so many in Europe had been displaced and dispossessed by the devastation of war during the 1790s. While so many of Kant's peers sought to read the faces of strangers and to calculate the threats posed by human movement, Kant posed a different consideration than the biometric dream of total identification. He asked what right a stranger has "entering foreign territory to be treated by its owner without hostility" and to what extent this right is a foundation that commits us to a project for peace rather than a system that enables violence and dehumanization (137). Although this call for a right of visitation and a right not to be treated as an enemy belongs to the particular time and place of Kant's Prussia and the transformation of Europe into a theater of war, it is supported by a simple yet complex idea that impels one toward peace: a law of universal hospitality. Beyond Kant's particular elaboration on how this law might work in practice and what it would or would not entail, I want to take note of how his elaboration on hospitality commits us to a particular operation of thought:

> We are speaking here, as in the previous articles, not of philanthropy, but of right; and in this sphere hospitality signifies the claim of a stranger entering foreign territory to be treated by its owner without hostility. The latter may send him away again, if this can be done without causing his death; but, so long as he conducts himself peaceably, he must not be treated as an enemy. It is not a right to be treated as a guest to which the stranger can lay claim—a special friendly compact on his behalf would be required to make him for a given time an actual inmate—but he has a right of visitation. This right to present themselves to society belongs to all mankind in virtue of our common right of possession on the surface of the earth on which, as it is a globe, we cannot be infinitely scattered, and must in the end reconcile ourselves to existence side by side: at the same time, originally no one individual had more right than another to live in any one particular spot. (137–138)

As another conceptual alternative to biometric thought, hospitality offers a way to think about human movement and strangers that is defined by reception rather than regulation and restriction and by reconciling ourselves to the reality of the existence of others at our side. If it commits us to living together—rather than imagining we can effectively separate one from another and from the world that sustains us—it does so because it acknowledges, rather than resolves, the complexity of that task. As Lindsay Anne Balfour notes, hospitality makes demands that are not just unanticipated; they are properly unidentifiable because they involve others and conditions that must be unknown, at least at first: "The extent to which the host is willing to extend a welcome to whoever or whatever arrives, before any indication of who or what that might be, is one of hospitality's most complex and enduring questions" (71–72). Given the challenges of an unconditional hospitality that is so different from the conditions that are attached to even the most well-meaning expressions of practical hospitality, there are numerous matters that must be carefully considered and which endanger an ethic of hospitality. Hospitality can effectively identify some to be at home while others are perceived to be strangers arriving. This scenario can readily transform a duty to another into a benevolent gift bestowed to another who is lucky to receive such charity. If it exists as a temporary measure, hospitality may not entail a long-term political strategy and thus will only ameliorate the worst effects of destructive economic and political practices. Hospitality may be both a necessary humanitarian response and the condition that forestalls more profound changes to the conditions that make life barely survivable for many. I raise the notion of hospitality not because it solves matters quickly and efficiently but because it does not: unlike biometric thought, it acknowledges such complex challenges that define shared existence on this planet and commits all of us to a project of thinking carefully about them. By refusing to identify another and by insisting that support is not conditional on identification, hospitality demands thought, care, reflection, and commitment.

Some might argue that all of this sounds like an argument bent on creating social adhesion at the cost of individual differences. I simply don't understand the premise of the complaint. In what sense are we, on a finite planet and almost universally affected by global capitalism, not powerfully attached and indeed dependent upon one another and the cultures we have created? The complaint contends that social adhesion, if we could imagine that it does not presently exist, will sacrifice meaningful individual existence. I am not convinced this is true, in large part because I see that individuality can and does exist as a product of social existence. Individuality is produced by social norms and conventions as a particular way of being in the world. Being responsive to social schemas of intelligibility that guide how we exist and how our existence can be recognized by ourselves and others involves understanding the particular ways in which any given person navigates these norms of recognition. And these norms can be understood to be utterly consequential even if they do not absolutely determine who one can be. My resistance to biometric thought is not a resistance to individual differences. In fact, I think biometric thought is profoundly incapable of acknowledging difference, except the narrow band of physical difference that it takes as the ground of identity.

If individual differences are going to matter and are going to help us recognize the singularity of each life on this planet, we have to be willing to think through social designations of difference. When biometric thought settles upon the physical body as the ground for identity, it does more than just recognize our singular existence. It recognizes our existence as independent from others by disconnecting that body from the world and from the social relations that make life possible. Social dependency is not a threat to difference but an acknowledgment that our singular existence depends upon social and environmental factors. A notion of individuality that fails to acknowledge this interdependency is suicidal in its refusal to comprehend what existence requires.

This complaint also fails to distinguish the claim of independence from the social experience of it, which is something else entirely. Biometric thought discovers that bodily life in itself is not sufficient to guarantee one's identity. Identity must be recorded and verified via measures that are external to the body itself. This discloses an important feature of ethical life, albeit in the service of a process that tends to efface this realization. The body is not enough. I cannot, according to biometrics, be myself on my own terms and in my own way. I must work with structures outside myself; thus, biometric practices reference a larger social world that impinges upon me and makes me possible as such. A body is at once mine and most identifiably mine on the basis that it possesses forms of intelligibility that I cannot confirm for myself and do not define. My body is mine because it exceeds my ability to know it. The very notion of mine-ness is itself a concept from a social world beyond me. This excess may be the opening to normative biometric designs that seek to monopolize the right to read the body and act exclusively as a force of social recognition of identity. But the social life of identity exists before and beyond biometric apprehensions of a bodily signature, and this can be the basis for recognizing that identity is not simply something grounded in the natural body.

Recognizing our shared social existence is not itself a solution to long-standing injustices or social inequalities that make a great many lives less survivable than others. Indeed, our shared social existence may drive some to verse—escaping violence, homophobia, racism, exploitation, sexism, environmental destruction, or economic abandonment—and thus it may be an expression of the problems some confront. Recognizing the social basis of life on this planet is, however, a means of making visible the conditions that regulate how so many live, conditions that biometric thought's priority on individuals seems scarcely prepared to acknowledge.

Derrida has noted that part of the power of hospitality is that it insists on "the public nature of public space" (*On Cosmopolitanism* 22). The force of this observation becomes clear when considered in relation to biometric thought's tendency to make human movement a private affair conducted by individuals in accordance with their individual desires and their encounters with particular laws. A public politics of hospitality can acknowledge that thinking about guests and hosts may mean entirely re-imagining the basis for the public in the first place. Biometric thought aids and abets a refusal in many parts of the Americas, for example, to acknowledge the historical crimes of colonialism and the dispossession of indigenous peoples by discouraging some from understanding themselves as guests within another's territory.

In my present home of Kelowna, British Columbia, a public hospitality must acknowledge the complicated histories of federal, provincial, and municipal aggression toward the Okanagan First Nations, histories of violence that continue legacies of racism and dispossession. Any idea of the public here must acknowledge the Syilx people's remarkable hospitality to those who, for generations, have sought to destroy and displace them and their culture. This means that when we use a language of hospitality that sees some as hosts to asylum-seeking foreigners, it is a mode of thought that can exclude an awareness of the guest-host relations already in operation here, erasing from public memory the historical theft of unceded indigenous land that has made some so prosperous.

What new ways of thinking about our obligations to one another and to the land might emerge from a public hospitality in which I come to understand my presence to be that of a guest rather than a host? Does it begin to seem strange for one guest to bar another or to have so little respect for our hosts and their rights? A public hospitality does not prescribe answers to these questions, but it does create conditions in which we can begin to act ethically by considering how we all live together on this planet and the complexity and responsibility to one another this entails.

As Peter Melville notes, if hospitality is "thought with any kind of rigour" it must include "its own failure ad infinitum," and thus the summons of hospitality "is endlessly open" (107). To be hospitable is to not only welcome someone at an appointed time for a particular purpose but also remain hospitable and open to the future and its uncertainties. This includes being attentive to the ways in which hospitality can be weaponized, especially when it is circumscribed by normative expectations about who will be welcomed, when it is deployed instrumentally as part of a means of displacing existing residents, or when it reduces ethics to expressions of benevolence.

Hospitality is a fragile thought, as this brief assessment of its nature and flexibility shows. It is capable of unsettling what we think of when we speak of guests and hosts, just as it is capable of unsettling itself. It can promise to transform the confidence with which we think we know who "we" are and who "they" are. Its power lies in this strangeness, too, this capacity to unmake what one knows and to differ from biometric thought's strident promises to make everything visible, ordered, named, identified.

What if we consider, as Wordsworth did, that identification could cease to be a matter of separating a body from its social and physical environment and more a matter of asking "How is it that you live"? My sense is that the full force of this question arrives not from the dialogue it establishes with another but from its scope and interest in a larger world that supports and fails to support one's existence. Speaking to one another matters and is an important alternative to inspecting another or projecting onto others what one wishes to find there. But dialogue—especially of the *Tell me about yourself* sort—will always remain constrained in advance by social codes and conventions that govern what one can say or can be heard to say. Dialogue should not be mistaken for a utopian vision of untroubled communication in which one shares the truth of oneself or hears the truth of another. In *Signs Preceding the End of the World*, Makina found herself sometimes heard above the din of racist presumptions that made her legible before she ever spoke, but not without considerable effort, and she also knew how the realities of a place transformed migrants and made their words and desires incomprehensible to those back home (20). To ask "How is it that you live?" is to hope for more than the value of conversation, then. It is a question that invites one to attend to the ways in which we exist as individual and social creatures whose lives depend on one another and on social and cultural conditions in order to unfold the way they do. Wordsworth's original query contains a certain respect and awe, and maybe a fatal curiosity, at another's survival. In order to lose some of these features that tend toward spectacle and which might even create yet another scene of inspection, perhaps the question is better simplified: "How do you live?"

How different might the world be if "Who's there?" was replaced by "How do you live?" At the very least, it means insisting first on the world itself as a powerful component of individual existence. It likely also invites an understanding of the ways in which one person's conditions of existence are shaped by the lives of others near and far. It seems to invite a moment of discussion on that basis and perhaps comparisons that invite a growing comprehension of the very different ways in which individuals are presently living and what makes such disparity possible. *Here is how I live. How do you live?* It certainly acknowledges that existence is conditioned by factors the individual does not choose but which nonetheless regulate how one lives and recognizes oneself and others. What kind of movement is possible under this question? Foregrounding living as something that one accomplishes, rather than the biological essence of working lungs and a beating heart, suggests a relationship to the conditions that make life possible and impossible. Perhaps identity becomes more contingent under the question "How do you live?" Identity begins to appear as more flexible, responsive, and defined by perseverance rather than as a permanent quality that transcends a social and natural environment.

"How do you live?" does not lose the normative horizon that structures identity under biometric thought. For example, how one lives may be subject to normative claims regarding how one ought to live, and this will mark as undesirable, intolerable, or inappropriate any number of legitimate ways of living. "How do you live?" may still invite a narrow understanding of oneself as an unattached individual responsible only for oneself and may even cultivate a sense that one lives despite the world, expressing a romantic rebelliousness that is all too ready to sacrifice social existence. Nor is one question better than another because it produces a narrative rather than an inspection. Identification might take a narrative form and still retain all of its power to foreclose important parts of who one is. Or, the occasion for telling the story might itself already make it impossible for one's narrative to be heard. This is one lesson of Thomas King's wonderful and heartbreaking story, "Borders," in which an indigenous woman and her child become stuck at the U.S.-Canadian border precisely because she insists on telling a true story of her nationality in which she identifies herself as Blackfoot. More, there is a long history of speaking with others that has meant constraining the terms within which one will be permitted to respond and what responses can be recognized as valid or heard and acknowledged at all. And as Wordsworth and King knew, it is always possible to be unwilling or unable to hear what another says in response to the question of "How do you live?"; thus, there is no promise that it will yield different results from asking "Who's there?"

But asking "How do you live?" may loosen up the punctual presence of identity enough to acknowledge the ways in which identity changes and depends upon circumstance, social bonds, the land and the air, culture, and history. The needs of the migrant, or how one understands oneself, in a new territory may be vastly different from what they were at home. How one understands the world and one-self may change with a new language and different customs. Biometric thought may not need to acknowledge such changes, preferring to chalk them up to personality rather than identity, but it is also clear that when it identifies some as illegal, for example, it makes a legal and social judgment that is absolute and implies the impossibility of another public existence for some migrants.

If the world is ordered according to the papered and the paperless, the legal and the illegal, and by regimes that verify identity and all the ways that these practices structure recognitions of social life, what kind of living is this that cannot tolerate difference and strangeness and uncertainty? What kind of memory is possible if we live in a world that refuses to remember the searching migrations, the creation of colonial empires, climate change, and the displacements of so many in times of conflict, famine, and economic deprivation that have transformed human history? What kind of responsibility is possible in a world that disqualifies individuals because they migrate when confronted with conditions that cannot be endured? What kind of identification is this that ignores the conditions involved in living with others? Jane Austen once wrote, "Seldom, very seldom does complete truth belong to any human disclosure; seldom can it happen that something is not a little disguised, or a little mistaken" (367). I find it helpful to remember that being human involves error and living with certain opacities and uncertainties and that not everything, or everyone, can be known or fully identified. The dreams of biometric thought reduce identity to embodied existence—whether by recording and verifying a gait, a thumbprint, or patterns on a retina—but perhaps we can remember rather than forget the losses collected as a result of seeing and knowing individuals only on these terms. Such a path of remembering begins with the recognition that who one is depends not on what can be measured and recorded biometrically but on what is beyond oneself, at the heart of oneself, a little disguised, and a little mistaken.

ACKNOWLEDGMENTS

This book was written largely in the unceded ancestral territory of the Syilx people, and I wish to acknowledge the inspiration that I have drawn from the hospitality they show toward migrants like me who have arrived here following paths of settlement forged by colonialism. I hope that some of the ideas contained here can speak to the appreciation I have for their survival and for the responsibility I have to break with that colonial past.

The creation of this book was thanks to support in large part from the Faculty of Creative and Critical Studies at the University of British Columbia, Okanagan campus. To have the time to research and write was invaluable.

Parts of this book appeared in print earlier, and I am grateful for the permission to reproduce some of that content here. Parts of chapters 1 and 3 first appeared in "Veiling and Other Fantasies of Visibility," which was published in *CR: The New Centennial Review* vol. 13, no. 3 (2013) and appears here by permission of Michigan State University Press. Part of chapter 5 first appeared in "Ethics in the Face of Terror: Shelley and Biometrics," which was published in *The Review of Education, Pedagogy & Cultural Studies* (www.informaworld.com), vol. 30, no. 3–4 (2008) and appears here by permission of Taylor & Francis Ltd. (www.tandfonline.com). A small part of chapter 5 first appeared in "Equiano's Refusal: Slavery, Suicide Bombing, and Negation," which was published in *European Romantic Review* (www.informaworld.com), vol. 27, no. 3 (2016) and appears here by permission of Taylor & Francis Ltd.

I wish to thank the artist Emerson Murray for his permission to use *Life Unfree* for the cover. I appreciate the permission of Universal News and Sport to reproduce their image of Banksy's graffiti in chapter 6.

I am especially grateful to Nicole Solano and Rutgers University Press for believing in this project and for finding thoughtful readers who understood what this book hopes to do. My thanks to those who reviewed the book; it is better for your interventions. I sincerely appreciate the work of Liz Asborno, who so effectively copyedited and typeset the manuscript. Thanks to Lisa Fedorak for creating the index. I am grateful, too, to all those at Rutgers University Press who helped to turn the manuscript into a finished book.

There are four who are closest to me, and they deserve far more recognition than I can bestow upon them here. To Sam, I wish to say that your friendship means the world to me. In what I have written here, I have tried as best I can to represent my thought to you, knowing that I may never have the pleasure of thinking with you in my classroom. To Zander and Charlie, you have both taught me more about how to live than I ever thought possible and in such very different ways. This book has benefited from our conversations and from our walks. To Joanna, you have changed me in more wonderful ways than I can ever thank you for. It has been such a joy to grow together in new ways, during the writing of this book and to discover just how much I still have to learn from you as a writer and a teacher. You have inspired many thoughts and feelings that guide much of what I have tried to say here regarding how individuals exist together. I have long known that those I am surrounded by are the very best parts of who I am, and you have each shown this to be true again and again.

WORKS CITED

ACLU. "Cracks in the System: 20 Years of the Unjust Federal Crack Cocaine Law." October 2006, https://www.aclu.org/other/cracks-system-20-years-unjust-federal-crack-cocaine-law.

Agamben, Giorgio. *Nudities*. Translated by David Kishik and Stefan Pedatella. Stanford UP, 2011.

Ajana, Btihaj. *Governing through Biometrics: The Biopolitics of Identity*. Palgrave Macmillan, 2013.

Austen, Jane. *Emma*. 1815, edited by Kristin Flieger Samuelian, Broadview, 2004.

Balfour, Lindsay Anne. *Hospitality in a Time of Terror: Strangers at the Gate*. Bucknell UP, 2018.

Balibar, Etienne. "The Borders of Europe." Translated by J. Swenson. *Cosmopolitics: Thinking and Feeling Beyond the Nation*, edited by Pheng Cheah and Bruce Robbins, U of Minnesota P, 1998, 216–232.

Barthes, Roland. *Mythologies*. Translated by Annette Lavers, The Noonday Press, 1972.

Bauman, Zygmunt, and David Lyon. *Liquid Surveillance: A Conversation*. Polity Press, 2013.

BBC News. "Polish and Italian MEPS Sanctioned for Hitler Salutes." 27 Oct. 2015, www.bbc.com/news/world-europe-34651255.

Bell, Charles. *The Anatomy and Philosophy of Expression as Connected with the Fine Arts*. London: Longman, Hurst, Rees and Orme, 1806.

Berlant, Lauren. *Cruel Optimism*. Duke UP, 2011.

"Biometric Identity Management System." 23 May 2016, United Nations High Commission for Refugees (UNHCR), www.unhcr.org/cgi-bin/texis/vtx/home/opendocPDFViewer.html?docid=550c304c9&query=bims.

Biometrics Institute. "Is Theft of a Biometric Possible?" FAQs, 16 Jan. 2018, www.biometrics institute.org/.

Biometrics Strategy and Forensic Services. 25 May 2018, UK House of Commons Science and Technology Committee, publications.parliament.uk/pa/cm201719/cmselect/cmsctech/800/800.pdf.

Black, Pamela, Stephen Porter, Alysha Baker, and Natasha Korva. "Uncovering the Secrets of the Human Face: The Role of the Face in Pro-Social and Forensic Contexts." *Facial Expressions: Dynamic Patterns, Impairments and Social Perceptions*, edited by Sandra T. Carter and Vaughn T. Bailey, Nova Science Publishers, 2012, 41–66.

Blake, William. "London." *Songs of Innocence and Experience*, The William Blake Archive, www.blakearchive.org/work/songsie. Accessed 13 May 2016.

Bradner, Eric, and Rene Marsh. "Acting TSA director reassigned after screeners failed tests to detect explosives, weapons." CNN, 2 June 2015, www.cnn.com/2015/06/01/politics/tsa-failed-undercover-airport-screening-tests/.

Brown, Wendy. *Walled States, Waning Sovereignty*. Zone Books, 2014.

Browne, Simone. "Race and Surveillance." *Routledge Handbook of Surveillance Studies*, edited by Kirstie Ball, Kevin D. Haggerty, and David Lyon, Routledge, 2012, 72–79.

Bujnowski, Rafal. *Flying Lessons*. Raster, 2004, raster.art.pl/gallery/artists/bujnowski/video.htm. Accessed 23 Sept. 2015.

Butler, Judith. *Frames of War: When Is Life Grievable?* Verso, 2009.

———. *Giving an Account of Oneself*. Fordham UP, 2005.

———. *Notes toward a Performative Theory of Assembly*. Harvard UP, 2015.

———. *Precarious Life: The Powers of Mourning and Violence*. Verso, 2004.

Butler, Judith, and Athena Athanasiou. *Dispossession: The Performative in the Political*. Polity, 2013.

Butler, Judith, and Gayatri Spivak. *Who Sings the Nation State?* Seagull Books, 2007.

Calvino, Italo. *If on a Winter's Night a Traveler*. Houghton Mifflin Harcourt, 1981.

"Canada Offers Leadership on the Syrian Refugee Crisis." Government of Canada, 24 Nov. 2015, www.canada.ca/en/immigration-refugees-citizenship/news/2015/11/canada-offers-leadership-on-the-syrian-refugee-crisis.html.

"Captured: People in Prison Drawing People Who Should Be." The Captured Project, thecapturedproject.com/. Accessed 28 May 2016.

Carretta, Vincent. *Equiano the African: Biography of a Self-Made Man*. U of Georgia P, 2005.

Chevalier-Skolnikoff, Suzanne. "Facial Expression of Emotion in Nonhuman Primates." *Darwin and Facial Expression: A Century of Research in Review*, edited by Paul Ekman, Academic Press, 1973, 11–89.

Cole, Simon A. *Suspect Identities: A History of Fingerprinting and Criminal Identification*. Harvard UP, 2002.

"Council removes Banksy artwork after complaints of racism." *The Guardian*. 1 Oct. 2014, www.theguardian.com/artanddesign/2014/oct/01/banksy-mural-clacton-racist.

Craven, Elizabeth. *Journey from the Crimea to Constantinople*. 1789, Gorgias Press, 2010.

Crivelli, Ernesto, Ruud De Mooij, and Michael Keen. "Base Erosion, Profit Shifting and Developing Countries." International Monetary Fund, May 2015, www.imf.org/external/pubs/ft/wp/2015/wp15118.pdf.

Darwin, Charles. *The Expression of the Emotions in Man and Animals*. 1872, Penguin, 2009.

Davidson, Cathy. "Olaudah Equiano, Written By Himself." *Novel: A Forum on Fiction*, vol. 40, no. 1–2, 2006–2007, 18–51.

Davis, Lennard J. *Enforcing Normalcy: Disability, Deafness, and the Body*. Verso, 1995.

De Man, Paul. *The Rhetoric of Romanticism*. Columbia UP, 1984.

Demir, Nilüfer. Photograph of Alan Kurdi. September 2015, Dogan News Agency.

Derrida, Jacques. "By Force of Law: The 'Mystical Foundation of Authority.'" *Deconstruction and the Possibility of Justice*, edited by Drucilla Cornell, Michel Rosenfeld, David Gray Carlson, Routledge, 1992, 3–67.

———.*On Cosmopolitanism*. Routledge, 2002.

———. *Paper Machine*. Translated by Rachel Bowlby, Stanford UP, 2005.

Dillman, Lisa. Translator's Note. *Signs Preceding the End of the World*, And Other Stories, 2015.

Ekman, Paul. "Darwin, Deception, and Facial Expression." *Annals of the New York Academy of Sciences*, vol. 1000, 2003, 205–221.

———. *Emotions Revealed*. New York: Henry Holt, 2003.

El-Enany, Nadine. "Aylan Kurdi: The Human Refugee." *Law Critique*, vol. 27, 2016, 13–15.

Elliot, Carl. *A Philosophical Disease: Bioethics, Culture, and Identity*. Routledge, 1999.

Equiano, Olaudah. *The Interesting Narrative of Olaudah Equiano, Written by Himself*. 1789, *The Interesting Narrative and Other Writings*, edited by Vincent Carretta, Penguin, 2003, pp. 1–306.

———. "Letter to *The Public Advertiser*." 1788, *The Interesting Narrative and Other Writings*, edited by Vincent Carretta, Penguin, 2003, 330–331.

Fagge, Nick. "Germany is 'overwhelmed' with false asylum seekers." *The Daily Mail Online*, 6 Nov. 2015, www.dailymail.co.uk/news/article-3305361/Germany-overwhelmed-false-asylum-seekers-Syrian-passports-forgery-experts-admit-t-spot-fakes.html.

Fanon, Frantz. "Algeria Unveiled." *Veil: Veiling, Representation and Contemporary Art*, edited by David A. Bailey and Gilane Tawadros, MIT Press, 2003, 72–85.

Faubert, Michelle. Introduction. *Mary, A Fiction and The Wrongs of Woman, or Maria*, Broadview Press, 2012, 11–50.

Finn, Jonathan. *Capturing the Criminal Image*. U of Minnesota P, 2009.

Foucault, Michel. *Discipline and Punish: The Birth of the Prison*. 1977, translated by Alan Sheridan, Random House, 1995.

———. "Foucault." Translated by Robert Hurley. *Aesthetics, Method, and Epistemology*, edited by James D. Faubion, The New Press, 1998, 459–463. Essential Works of Foucault 1954–1984.

———. *The History of Sexuality, Volume 1: An Introduction*. 1978, translated by Robert Hurley, Random House, 1990.

———. "Life: Experience and Science." Translated by Robert Hurley. *Aesthetics, Method, and Epistemology*, edited by James D. Faubion, The New Press, 1998, 465–478. Essential Works of Foucault 1934 1984.

———. *Security, Territory, Population: Lectures at the Collège de France*. Translated by Graham Burchell, edited by Michel Senellart, Palgrave Macmillan, 2007.

———. "Structuralism and Poststructuralism." Translated by Jeremy Harding. *Aesthetics, Method, and Epistemology*, edited by James D. Faubion, The New Press, 1998, 433–458. Essential Works of Foucault 1954–1984.

Gaiser, Sara. "San Francisco Supervisor Aaron Peskin Proposes Citywide Ban on Facial Recognition Technology." *The San Francisco Examiner*, 29 Jan. 2019, www.sfexaminer.com /peskin-proposes-citywide-ban-facial-recognition-technology/.

Galperin, William H. *The Return of the Visible in British Romanticism*. The Johns Hopkins UP, 1993.

"German Parliament Approves IDs for Refugees." Deutsche Welle, 14 Jan. 2016, www.dw.com /en/german-parliament-approves-ids-for-refugees/a-18981451.

Groebner, Valentin. *Who Are You? Identification, Deception, and Surveillance in Early Modern Europe*. Translated by Mark Kyburz and John Peck, Zone Books, 2007.

Herrera, Yuri. *Signs Preceding the End of the World*. Translated by Lisa Dillman, And Other Stories, 2015.

Higgs, Edward. *Identifying the English*. Continuum, 2011.

Hill, Sid. "My Six Nation Haudenosaunee Passport Is Not a 'Fantasy Document.'" *The Guardian*, 30 Oct. 2015, www.theguardian.com/commentisfree/2015/oct/30/my-six-nation -haudenosaunee-passport-not-fantasy-document-indigenous-nations.

"The Hon. Chris Alexander and the Hon. Kellie Leitch Announce Measures to Stop Barbaric Cultural Practices Against Women and Girls." Conservative Party of Canada, 2 Oct. 2015, www.conservative.ca/the-hon-chris-alexander-and-the-hon-kellie-leitch-announce -measures-to-stop-barbaric-cultural-practices-against-women-and-girls-2/. Accessed 17 Oct. 2015.

Jacobsen, Katja Lindskov. "On Humanitarian Refugee Biometrics and New Forms of Intervention." *Journal of Intervention and Statebuilding*, vol. 11, no. 4, 2017, 529–551.

———. *The Politics of Humanitarian Technology*. Routledge, 2015.

Jain, Jasbir. "Patriarchy, and the Tropical Sun: Womanhood in India." *The Veil: Women Writers on Its History, Lore, and Politics*, edited by Jennifer Heath, U of California P, 2008, 231–247.

Jones, Julia, and Eve Bower. "American Deaths in Terrorism vs. Gun Violence in One Graph." CNN, 30 Dec. 2015, www.cnn.com/2015/10/02/us/oregon-shooting-terrorism-gun -violence/.

Juengel, Scott. "Godwin, Lavater, and the Pleasures of Surface." *Studies in Romanticism*, vol. 35, no. 1, 1996, 73–97.

Kant, Immanuel. "An Answer to the Question: What is Enlightenment." Translated by David L. Colclasure. *Toward Perpetual Peace and Other Writings on Politics, Peace, and History*, edited by Pauline Kleingeld, Yale University Press, 2006, 17–23.

———. *Perpetual Peace: A Philosophical Essay*. Translated by M. Campbell Smith, Garland, 1972.

Kenney, Jason. "On the Value of Canadian Citizenship." Government of Canada, 12 Dec. 2011, https://www.canada.ca/en/immigration-refugees-citizenship/news/archives/speeches-2011/jason-kenney-minister-2011-12-12.html.

Khan, Shahnaz. "Afghan Women: The Limits of Colonial Rescue." *Feminism and War: Confronting US Imperialism*, edited by Robin Lee Riley, Chandra Talpade Mohanty, and Minnie Bruce Pratt, Zed Books, 2008, 161–178.

Kilpatrick, Kate. "U.S.-Mexico Border Wreaks Havoc on Lives of an Indigenous Desert Tribe." Aljazeera, 24 May 2014, america.aljazeera.com/articles/2014/5/25/us-mexico-borderwreakshavocwithlivesofanindigenousdesertpeople.html.

King, Thomas. "Borders." *One Good Story That One*, U of Minnesota P, 2013, 131–151.

Laurent, Olivier. "What the Image of Aylan Kurdi Says About the Power of Photography." *Time*, 4 Sept. 2015, time.com/4022765/aylan-kurdi-photo/.

Lavater, Johann Caspar. *Essays on Physiognomy; For the Promotion of the Knowledge and the Love of Mankind*. Translated by Thomas Holcroft, G.G.J. and J. Robinson, 1789. 3 vols.

Laymon, Kiese. "My Vassar College ID Makes Everything Ok." *Gawker*, 29 Nov. 2014, gawker.com/my-vassar-college-faculty-id-makes-everything-ok-1664133077.

Lee, Justin. "Indian Airport Implements Biometric Technology for Paperless E-Boarding." Biometric Update, 14 Aug. 2015, www.biometricupdate.com/201508/indian-airport-implements-biometric-technology-for-paperless-e-boarding.

Levinas, Emmanuel. "Responsibility and Substitution." Translated by Maureen Gedney. *Is It Righteous To Be?*, edited by Jill Robbins, Stanford UP, 2001, 228–233.

Lish, Atticus. *Preparations for the Next Life*. Tyrant Books, 2014.

Lloyd, Martin. *The Passport: The History of Man's Most Travelled Document*. Sutton Publishing, 2003.

Lyon, David, Kevin D. Haggerty, and Kirstie Ball. "Introducing Surveillance Studies." *Routledge Handbook of Surveillance Studies*, edited by Kirstie Ball, Kevin D. Haggerty, and David Lyon, Routledge, 2012, 1–11.

Magnet, Shoshana Amielle. *When Biometrics Fail: Gender, Race, and the Technology of Identity*. Duke UP, 2011.

Marks, Deborah. *Disability: Controversial Debates and Psychosocial Perspectives*. Routledge, 1999.

Marx, Gary T. "'Your Papers Please': Personal and Professional Encounters with Surveillance." *Routledge Handbook of Surveillance Studies*, edited by Kirstie Ball, Kevin D. Haggerty, and David Lyon, Routledge, 2012, xx–xxxi.

Masood, Maliha. "On the Road: Travels with My Hijab." *The Veil: Women Writers on Its History, Lore, and Politics*, edited by Jennifer Heath, U of California P, 2008, 213–227.

McGann, Jerome. "The Beauty of the Medusa: A Study in Literary Iconography." *Studies in Romanticism*, vol. 11, 1972, 3–25.

Melville, Peter. *Romantic Hospitality and the Resistance to Accommodation*. Laurier UP, 2007.

Miller, J. Hillis. *For Derrida*. Fordham UP, 2009.

Monahan, Torin. *Surveillance in the Time of Insecurity*. Rutgers UP, 2010.

Mordini, Emilio. "Biometrics." *Handbook of Global Bioethics*, edited by Henk A. M. J. ten Have and Bert Gordijn, Springer, 2014, pp. 504–526.

Muller, Benjamin J. "Securing the Political Imagination: Popular Culture, the Security Dispositif and the Biometric State." *Security Dialogue*, vol. 39, no. 2–3, 2008, 199–220.

Nanavati, Samir, Michel Thieme, and Raj Nanavati. *Biometrics: Identity Verification in a Networked World*. John Wiley and Sons, 2002.

National Research Council. *Biometric Recognition: Challenges and Opportunities*. Edited by Joseph N. Pato and Lynette I. Millett, National Academies Press, 2010.

Ngũgĩ wa Thiong'o. *Decolonizing the Mind: The Politics of Language in African Literature.* East African Educational Publishers, 1986.

Nguyen, Mimi. *The Gift of Freedom: War, Debt, and Other Refugee Passages.* Duke UP, 2012.

Nixon, Mark S., Tieniu Tan, and Rama Chellappa. *Human Identification Based on Gait.* Springer, 2006.

"No Wall." Tohono O'odham Nation, www.tonation-nsn.gov/nowall/. Accessed 13 Nov. 2017.

Nussbaum, Felicity. *The Limits of the Human: Fictions of Anomaly, Race, and Gender in the Long Eighteenth Century.* Cambridge UP, 2003.

Ojebode, Ayobami. "Alan Kurdi, Deaths in the Desert and Failed Migrants' Processing of Dystopic Images on Social Media." *Crossings: Journal of Migration and Culture,* vol. 8, no. 2, 2017, 115–130.

Oxford English Dictionary (OED). "Biometrics." www.oed.com/view/Entry/273387?redirected From=biometrics#eid.

"Panama Papers Expose Human Cost of Global Tax Avoidance." *The Current,* CBC Radio, 11 Apr. 2016, www.cbc.ca/radio/thecurrent/the-current-for-april-11-2016–1.3529740/apr-11 -2016-episode-transcript-1.3531071#segment2.

Pearl, Sharrona. *About Faces: Physiognomy in Nineteenth-Century Britain.* Harvard UP, 2010.

Pearson, Karl. *National Life from the Standpoint of Science.* 2nd ed., Adam and Charles Black, 1905.

The Pocket Lavater, or, The Science of Physiognomy. New York: Van Winkle and Wiley, 1817.

Pugliese, Joseph. *Biometrics: Bodies, Technologies, Biopolitics.* Routledge, 2010.

Redfield, Marc. *The Politics of Aesthetics: Nationalism, Gender, Romanticism.* Stanford UP, 2003.

Rediker, Marcus. *The Slave Ship: A Human History.* Penguin, 2007.

Report of the Special Rapporteur on the human rights of migrants, François Crépeau, on his mission to Qatar. United Nations, 23 Apr. 2014, www.ohchr.org/Documents/Issues/SRMigrants/A -HRC-26-35-Add1_en.pdf.

Roache, Trina. "Mi'kmaq Hereditary Chief Calls on Indigenous Leaders to Press Ottawa on Syrian Refugees." Aboriginal Peoples Television Network, 21 Sept. 2015, aptn.ca/news/2015/09/21 /mikmaq-hereditary-chief-calls-on-indigenous-leaders-to-press-ottawa-on-syrian-refugees/.

Said, Edward. *Orientalism.* Penguin, 1977.

Salter, Mark. "Passports, Mobility, and Security: How Smart Can the Border Be?" *International Studies Perspectives,* vol. 5, no. 1, 2004, 71–91.

Santos, Fernanda. "Border Wall Would Cleave Tribe, and Its Connection to Ancestral Land." *The New York Times,* 20 Feb. 2017, www.tonation-nsn.gov/wp-content/uploads/2017/02 /Border-Wall-Would-Cleave-Tribe-and-Its-Connection-to-Ancestral-Land.pdf.

Scott, Joan Wallach. *The Politics of the Veil.* Princeton UP, 2007.

Sedgwick, Eve Kosofsky. *Epistemology of the Closet.* U of California P, 1990.

———. *Touching Feeling: Affect, Pedagogy, Performativity.* Duke UP, 2003.

Serlin, David. "Ready for Inspection: An Interview with Valentin Groebner." *Cabinet Magazine,* vol. 22, Summer 2006, www.cabinetmagazine.org/issues/22/serlin.php.

Shakespeare, William. *Hamlet. The Riverside Shakespeare,* edited by G. Blakemore Evans, Houghton Mifflin, 1974.

Shelley, Mary. *Frankenstein.* 1818, Broadview Press, 2012.

Shelley, Percy Bysshe. *The Complete Poetical Works of Percy Bysshe Shelley.* Edited by Thomas Hutchinson, Oxford UP, 1967.

———. *The Letters of Percy Bysshe Shelley.* Edited by Frederick L. Jones, vol. 2, Clarendon Press, 1964.

Simpson, David. *9/11: The Culture of Commemoration.* U of Chicago P, 2006.

———. *Romanticism and the Question of the Stranger.* U of Chicago P, 2013.

————. *Wordsworth, Commodification and Social Concern: The Poetics of Modernity.* Cambridge UP, 2009.

Stewart, Robert (Lord Castlereagh). "Speech delivered in the House of Commons, 16 May 1821." *The Parliamentary Debates*, vol. 5, T. C. Hansard, 1822.

Tan, Vivian. "UNHCR's New Biometrics System Helps Verify 110,000 Myanmar Refugees in Thailand." United Nations High Commission on Refugees, 30 June 2015, www.unhcr.org /55926d646.html.

Torpey, John. *The Invention of the Passport: Surveillance, Citizenship and the State.* Cambridge UP, 2000.

Turhan, Filiz. *The Other Empire: British Romantic Writings about the Ottoman Empire.* Routledge, 2003.

"Twin Souls Project." Change One Life, odnolico.changeonelife.ru/en/about. Accessed 18 Jan. 2018.

"The Ugly Side of the Beautiful Game." Amnesty International, 2016, www.amnesty.org /download/Documents/MDE2235482016ENGLISH.PDF.

Universal Declaration of Bioethics and Human Rights. United Nations Education, Scientific and Cultural Organization (UNESCO), 19 Oct. 2005, portal.unesco.org/en/ev.php-URL _ID=31058&URL_DO=DO_TOPIC&URL_SECTION=201.html.

Universal Declaration of Human Rights. United Nations, 1948, www.un.org/en/universal -declaration-human-rights/.

Van der Ploeg, Irma. *The Machine Readable Body: Essays on Biometrics and the Informatization of the Body.* Shaker Publishing, 2005.

Vaughan-Williams, Nick. *Border Politics: The Limits of Sovereign Power.* Edinburgh UP, 2009.

Viswanathan, Guari. *Masks of Conquest: Literary Study and British Rule in India.* Columbia UP, 1989.

Wacks, Raymond. *Privacy: A Very Short Introduction.* Oxford UP, 2010.

Waddell, Kaveh. "Police Can Force You to Use Your Fingerprint to Unlock Your Phone." *The Atlantic*, 3 May 2016, www.theatlantic.com/technology/archive/2016/05/iphone-fingerprint -search-warrant/480861/.

Wallace, Rebecca. "Contextualizing the Crisis: The Framing of Syrian Refugees in Canadian Print Media." *Canadian Journal of Political Science*, vol. 51, no. 2, 2018, 207–231.

Walters, William. "Rezoning the Global: Technological Zones, Technological Work and the (un-)Making of Biometric Borders." *The Contested Politics of Mobility: Borderzones and Irregularity*, edited by Vicki Squire, Routledge, 2011, 51–73.

Wolfe, Carey. *What Is Posthumanism?* U of Minnesota P, 2010.

Wordsworth, William. *Poems, in Two Volumes, and Other Poems, 1800–1807.* Edited by Jared Curtis, Cornell UP, 1983.

Worth, Katie. "Can Biometrics Solve the Refugee Debate?" *Frontline*, PBS, 2 Dec. 2015, www .pbs.org/wgbh/frontline/article/can-biometrics-solve-the-refugee-debate/.

Youngquist, Paul. "The Afro Futurism of DJ Vassa." *European Romantic Review*, vol. 16, no. 2, 2005, 181–192.

Zahedi, Ashraf. "Concealing and Revealing Female Hair: Veiling Dynamics in Contemporary Iran." *The Veil: Women Writers on Its History, Lore, and Politics*, edited by Jennifer Heath, U of California P, 2008, 250–265.

Žižek, Slavoj. "Da Capo Senza Fine." *Contingency, Hegemony, Universality: Contemporary Dialogues on the Left*, by Judith Butler, Ernesto Laclau, and Slavoj Žižek, Verso, 2000.

————. *Violence.* Picador, 2008.

INDEX

Page numbers in italics indicate photographs.

ABOUT THE AUTHOR

GEORGE C. GRINNELL is an associate professor in the department of English and Cultural Studies at the University of British Columbia, Okanagan campus, where he teaches critical theory, punk culture, and literature of the Romantic era. His research is consistently concerned with how life is conditioned by social and cultural norms and forms of legibility and illegibility. He is the author of *The Age of Hypochondria: Interpreting Romantic Health and Illness* (2010) as well as numerous book chapters and articles in journals such as *Studies in Romanticism, CR: The New Centennial Review,* and *European Romantic Review.* He lives in Kelowna, BC, and can often be found in the mountains with a Carolina dog named Charlie nearby.

Printed in the United States
By Bookmasters